Science
for the
70's

Science for the 70's

BOOK TWO

A. J. Mee

Patricia Boyd
Principal Biology Teacher
Broughton Secondary School, Edinburgh

David Ritchie
Deputy Headmaster and Principal Physics Teacher
Balwearie School, Kirkcaldy

Heinemann Educational Books
London and Edinburgh

Heinemann Educational Books Ltd

London Edinburgh Melbourne Auckland Toronto
Hong Kong Singapore Kuala Lumpur
Ibadan Nairobi Johannesburg
Lusaka New Delhi

ISBN 0 435 57576 7
© A. J. Mee, P. Boyd, and D. Ritchie 1971

First published 1971
Reprinted 1972 (twice)
Second Edition in full colour 1974
Reprinted 1975

Published by Heinemann Educational Books Ltd
48 Charles Street, London W1X 8AH

Printed in Great Britain by
Jarrold and Sons Ltd, Norwich

A Note to the Teacher

Science for the Seventies, a series of two pupils' books with Teachers' Guides, comprises a complete experimental Integrated Science Course for the first two or three years of secondary education. It follows the order of the Scottish Integrated Science syllabus laid down in *Curriculum Paper Number 7: Science in General Education* (H.M.S.O.).

It can be used in conjunction with the Scottish Secondary Science Working Party's **Science Worksheets** (Heinemann Educational Books) or independently.

The *Teachers' Guide* for each book function both with **Science for the Seventies** and the **Science Worksheets.** Teachers are recommended to consult the *Guides* in order to make full use of the pupils' books.

A.J.M.
P.B.
1971   D.R.

A Note for Pupils only

THIS BOOK has been written for your enjoyment. You may think this is a strange thing to say about a school text-book; an adventure story, yes – but not a school book! But, you see, we do not believe that any subject, least of all science, should be dry and distasteful. Let us assure you that this is not just another old-fashioned kind of school book, giving you only the facts about the subject which you must learn – or else!

We do not believe in telling you everything. Being intelligent young people there is much that you can work out, and so quite often we have left you to do the thinking, and we are sure you will enjoy doing that. Science cannot really be learnt from a book anyway, but only by carrying out experiments for yourselves. This book is full of them, and we hope you will be able to do most, if not all, of these experiments either in your school laboratory or at home.

In order that you may be sure that you really have learnt something we have put a summary at the end of each unit showing the points you should know about and remember.

A. J. Mee
Patricia Boyd
1971   *David Ritchie*

Contents

Acknowledgements

Acknowledgements for permission to publish photographs are due as follows:

Aerofilms, 12.2, 12.3
Air Ministry (Crown Copyright), 10.1
Airviews Ltd, 14.1
Associated Press, 14.2
Bell Lines, 13.2 *top*
Brian Bracegirdle, 13.25, 14.12, 14.17
British Hovercraft Corporation Ltd, 13.8
British Rail, 15.36
British Steel Corporation, 12.9, 12.10
Bruce Coleman Ltd, (Russ Kinne) 9.17 *top*, (Sven Gillsater) 9.17 *bottom*, (Leonard Lee Rue) 13.33, (Donald Paterson) 13.33 *bottom*
Costain, 13.2 *top*, 13.24
Church Missionary Society, 14.2
Stephen Dalton, 13.26 *top left*, 13.26 *centre left*, 13.26 *right* (centipede), 13.26 *bottom right*
Do It Yourself Magazine, 9.3, 9.9
European Books Ltd, 14.11 *bottom*

Fisons Ltd, 10.10
Ginge Raadvad, 13.1 *top*
Hestair Sherpa Ltd, 13.10
ICI Ltd, 10.7
Institute of Geological Sciences (Crown Copyright), 12.1, 12.4, 12.6, 12.16, 12.17, 12.18, 12.23
London Brick Co. Ltd, 12.12
Natural History Photographic Agency, 9.17 *centre*, 13.26 *bottom left*, 13.26 *right* (prawn), 14.8
National Youth Orchestra, 11.21, 11.24
Paul Popper Ltd, 13.1 *bottom right*
Radio Times Hulton Picture Library, 11.13, 15.25
St Mary's Hospital Medical School, 11.29, 13.27
Scottish Tourist Board, 12.11
Shell Ltd, 12.14
B. W. Smith Photography Ltd, 13.8
South of Scotland Electricity Board, 15.35
United States Information Service, 13.20

Front cover: Volcanic eruption in Hekla, Iceland. *Picturepoint Ltd*

Back cover: Giraffe and zebra in the Nakuru District in Kenya. *Bruce Coleman Ltd (Jane Burton)*

Unit Nine
Heat Flow

9.1 PARTICLES AND ENERGY

We must start by reminding you of something you learnt in Book 1. Do you remember finding out what we believe matter to be made up of, and how the different states of matter – solid, liquid, and gas – fit into the picture?

Just what comes into your mind when you think of solids, liquids, and gases? Can you remember how we imagined particles to be arranged in the three different states? In which state do the particles have most energy?

We believe that a gas is made up of particles which are well spread out, because this idea explains a lot of the properties of gases – the fact that they can move easily from one place to another, that they are light, and that they are

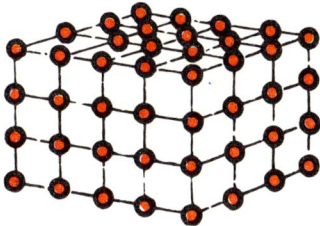

solid state

Fig. 9.1

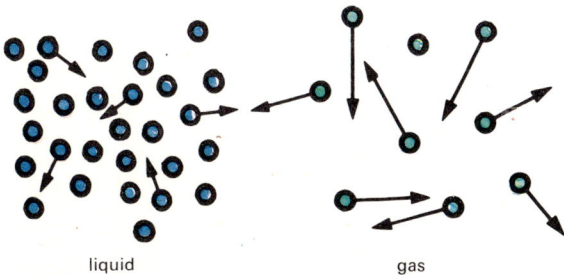

liquid   gas

easily compressed. Each particle in a gas has a considerable amount of movement energy so that they are all travelling quite quickly. The particles in air under ordinary room conditions are moving with something like the speed of a rifle bullet. They collide with each other and with their surroundings.

In a liquid, the particles are much closer together. Although they pull on each other they are all able to flow about, but with less energy and at slower speeds than gas particles.

Particles in a solid are not able to flow about, and are held in place in ranks and files or as part of some definite pattern. Such particles cannot change their places, but when given energy they vibrate from side to side. The more energy they have the more they vibrate. Because of the increased vibration and movement energy of the particles when they are heated, they take up more room. Can you remember what we called this effect?

9.2 HOW HEAT ENERGY TRAVELS

In this unit we are going to find out how heat energy travels in different substances. If our ideas about particles are correct we should be able to explain what happens by using them.

Experiment 9.1
Heating a metal rod

The metal rod in the diagram has some tacks or rivets stuck to it at different distances along the rod by wax or vaseline. If the wax or vaseline gets hot what do you think will happen to the rivets?

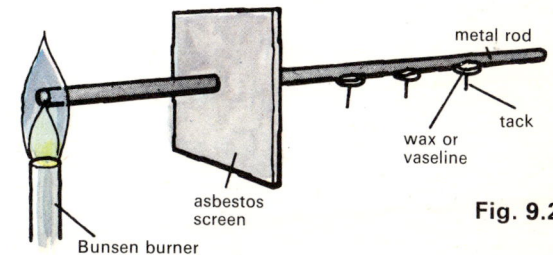

metal rod

tack

wax or
vaseline

asbestos
screen

Bunsen burner

Fig. 9.2

Heat the rod and watch the rivets. Which one falls off first, which second, and so on? What does this order tell you?

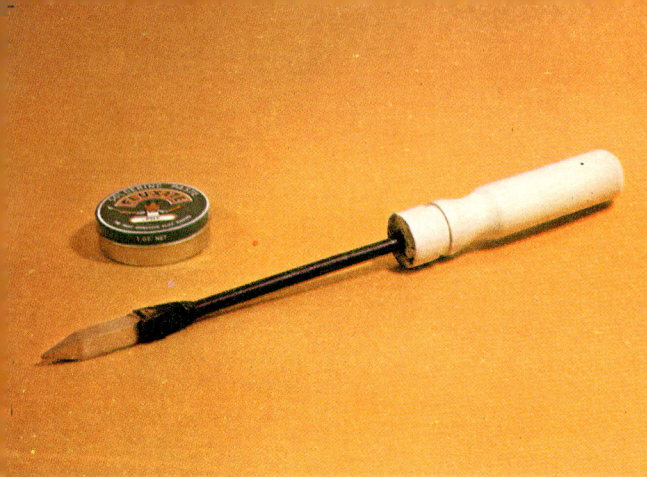

Fig. 9.3 A soldering iron. How does it work? Why does it have a wooden handle?

We see that the heat energy has travelled along the rod and has been passed on from one section of the rod to the next.

Does our particle theory explain this? When we heat the rod what do we imagine the particles will do? As they are in the solid state they are held in place but vibrate in their positions. If we continue to supply them with energy they vibrate more and more. They move so far from side to side that they affect their neighbours, and these in turn vibrate more energetically and pass energy on. In this way the energy will pass along the metal rod. We call this method of transferring heat from one particle to another **conduction**.

Experiment 9.2
Heating a liquid

Put a single crystal of potassium permanganate at the bottom of a beaker of water, as near to one edge as you can. What will happen to the crystal in the water?

Now put a small flame underneath the crystal. What happens? What do you notice about the direction in which the solution moves? What happens when it gets to the top? Will it be only particles of the crystal which move in this way? There is no reason to think so. Although we cannot see the water particles they must be moving in the same way as the permanganate ones.

Why should the particles move in this direction? What will happen to the spacing of the particles in a liquid as it is heated? What effect will this have on the density of the liquid? If the density becomes less where would you expect to find this less dense liquid — at the bottom or the top of the liquid?

We hope you can answer these questions for yourself, but if you cannot here are some hints. When a liquid is heated its particles move more quickly and spread out thus becoming less densely packed. We have learnt in Unit 4 that less dense liquids float on more dense ones.

When the hot liquid gets to the top it cools. So what should happen to it then, if our particle theory is correct? Is this what actually took place?

You will see that the hot liquid, when it rises to the top, takes its heat with it, so this is another way in which heat energy can be taken from one place to another. This effect of a liquid rising when heated and sinking when cooled, so that the liquid circulates, is called **convection**.

Why could convection not take place in a solid? Can it take place in a gas?

Experiment 9.3
Heating by waves

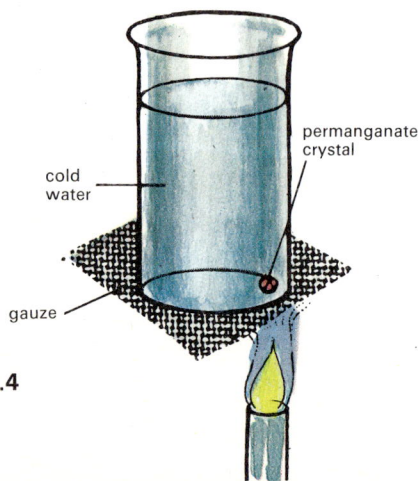

Fig. 9.5

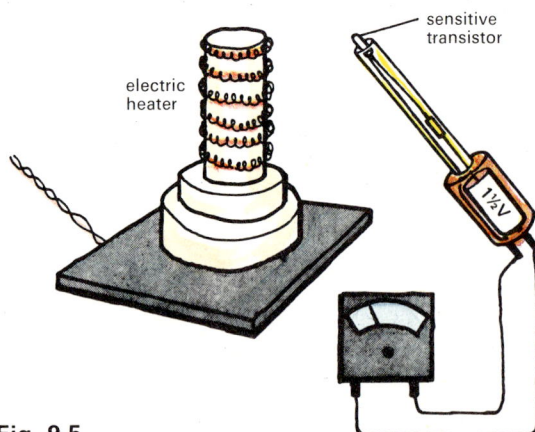

Fig. 9.4

This experiment uses a heating element which is attached to the electricity mains and should not be touched.

When the coil of wire in the electric heater is switched on, place your hand about 25 cm from the apparatus, first above, then alongside, and finally below. How does your hand feel in each case?

As you will find out for yourself a little later, the heat cannot be travelling to your hand entirely by conduction or convection. Put a photo-transistor connected to a sensitive ammeter (a micro-ammeter) in place of your hand in the positions mentioned. Is there any movement of the pointer of the micro-ammeter? The only kind of energy that can affect the transistor in this way is heat energy which has travelled in waves. We call this kind of heat energy 'radiant' energy, radiation, or infrared energy, and we believe that it travels in invisible waves because, as you will learn later on, it behaves just as waves do.

9.3 MORE EXPERIMENTS ON HEAT FLOW

The experiments mentioned below are arranged round the laboratory, and you will move from one to the next when your teacher tells you to do so. It does not matter in which order you do them, but be careful to make a note of what you observe, either in your note book or on the worksheet.

Experiment 9.4
Comparing the rates at which heat is conducted by different solids

Four different materials are provided. To be quite fair about comparing them, what should have been arranged about the materials? Look

at their lengths and thicknesses. Do these agree with what you thought?

Remember that you must start heating the ends of all the rods at the same time. Arrange your apparatus in such a way as to do this.

We have already seen one way of finding out how far along the rod the heat has been conducted. Can you suggest another method? There are many ways you could use, but here are two hints. What about sliding match heads, or pieces of 'heat sensitive' paper, along the rods? How would these tell you how far the heat has reached? If you decide to use matches make sure you hold on to the match stick.

Put the conductors in order starting with the best and finishing with the worst.

Experiment 9.5
Does heat travel through water equally well both upwards and downwards?

Pour cold water in the test-tubes which should be then clamped at an angle (Fig. 9.7).

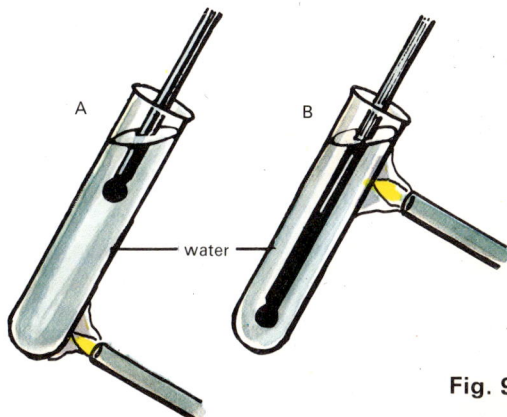

water

Fig. 9.7

Make sure that the outsides of the tubes are dry. You and your partner should now each take the temperature of the water in the test-tubes A and B as shown. Make sure that the bulbs of the thermometers are at different levels. Now each start to heat your tubes gently at the points shown. After about 20 seconds make a note of the temperature readings again;

	Tube A	Tube B
Temperature at start		
Temperature after 20 s		
Rise in temperature		

Fig. 9.6

do not move the thermometers from their positions during this time. Now remove the Bunsens, and allow the tubes to cool.

Did both thermometers show the same change in temperature? In which direction has the heat travelled? If very little heat has travelled downwards in tube B, what does this tell us about water as a conductor of heat?

Experiment 9.6
Is air a good conductor of heat?

This is very similar to the last experiment. First take the temperatures of the air in tubes A and B (Fig 9.8), then heat gently where shown.

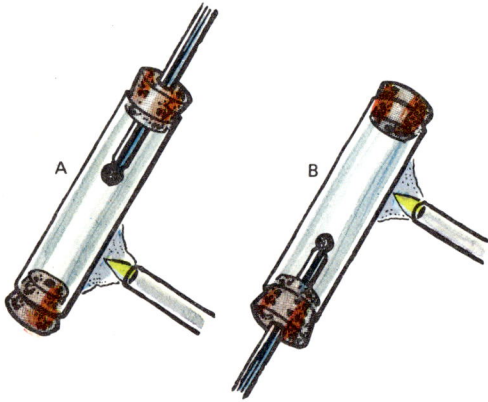

Fig. 9.8

Fig. 9.9 Insulation of roof, cavity walls, and pipes. What is the similarity between the various insulating materials used?

After about 10 seconds, again note the temperatures and remove the Bunsens. Make sure both readings are taken at the same time.

What changes of temperature have occurred during this time? In which direction has the heated air moved? Does this tell you if air is a good or bad conductor of heat?

Can you suggest why there should be a hole in one of the stoppers?

When you have completed Experiments 9.4, 9.5, and 9.6 you should be able to put water, air, and metals in order as conductors of heat. Which of them is the poorest?

Bad conductors of heat are often called insulators. You have come across this term before in Unit 7, where bad conductors of **electricity** were also called insulators. Very often materials such as wool, straw, layers of newspaper, feathers,

and expanded polystyrene are used as insulators to prevent heat from escaping and so to prevent cooling. Of course, they work the other way round too; they will prevent cold things from getting hot.

If you defrost the refrigerator you may be told to take foods from the cooling compartment and wrap them in layers of newspaper until the refrigerator is ready again. Why is that?

When you buy ice-cream it is sometimes put into a special container. What is the container made of, and why is it used?

The hot-water tank in your house will probably be 'lagged' to stop it from losing heat. Find out what material has been used. If you cannot see what it is, go to a 'do-it-yourself' shop and ask what they sell for this job.

Can you think what all the bad conductors listed above have in common, and what helps them to be such poor conductors?

Have you noticed that bad conductors of heat are also bad conductors of electricity? This makes us think that there must be some connection between the two.

Experiment 9.7
What makes the propeller move?

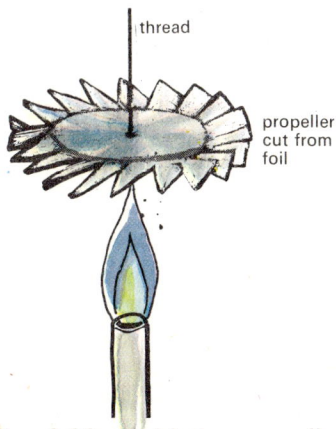

Fig. 9.10

Look at Fig. 9.10. Hold the propeller over a Bunsen flame. Be careful not to allow the material to touch the flame itself. Does the propeller move? Can you explain why? If you are in doubt, try blowing upwards from the position where the Bunsen was. What does this show about the air heated by the Bunsen?

This should remind you of Experiment 9.2. What method of heat transfer do we call this? The air particles behave in a similar way to the liquid particles in that experiment.

Experiment 9.8
Which cools faster?

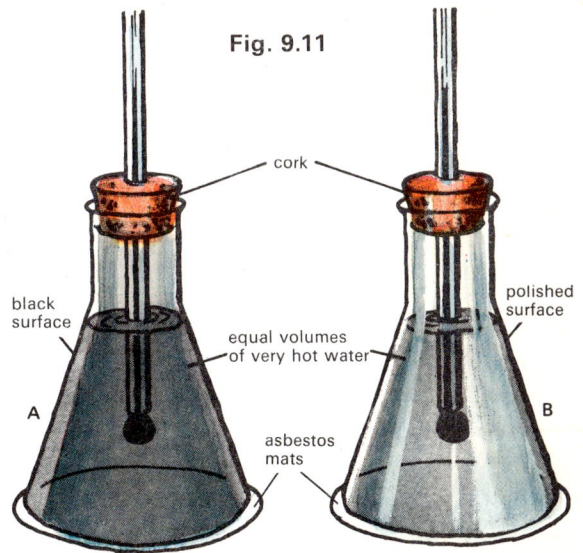

Fig. 9.11

In this experiment fill each flask (Fig. 9.11) to the same level with almost boiling water. We are going to find out whether the water in the black or the polished flask gives away its heat faster and so cools more quickly.

Each flask should be placed on an asbestos mat. This material is an insulator, and not much heat should be conducted into the bench.

Why should we have equal volumes of water in each flask? If one flask had only a little water in it do you think it would remain hot long?

When the corks and thermometers are fitted into the necks of the flasks, take the highest temperature of each flask, and then note the temperature of each flask every minute for 5 minutes. Which flask has cooled quicker and

Time in minutes	1	2	3	4	5
Flask A					
Flask B					

so given away more heat? As air is a bad conductor, do you think most of the heat has been lost by convection or by radiation? It may help you if you remember Experiment 9.3.

Most heat here has, in fact, been lost by radiation. We always find that dull dark surfaces radiate away more heat energy than bright polished ones.

Experiment 9.9
Reflecting heat

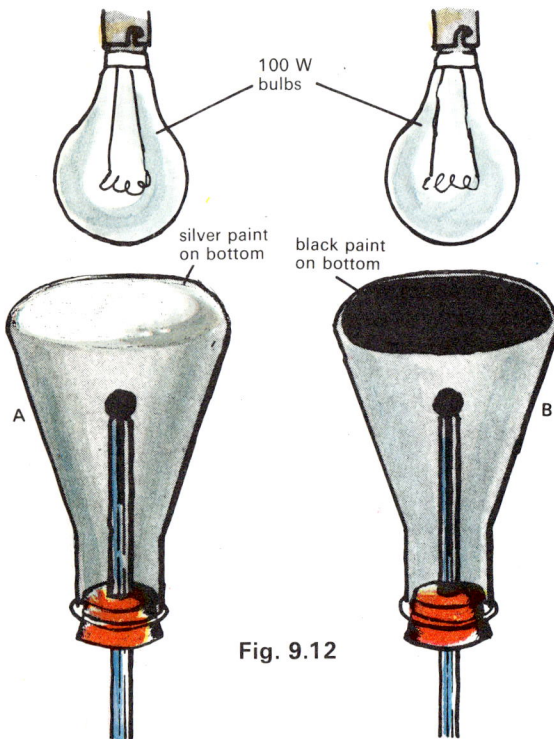

100 W bulbs

silver paint on bottom

black paint on bottom

A   B

Fig. 9.12

In the apparatus shown in Fig. 9.12, note the temperature shown by the thermometers. Now switch on both bulbs at the same time, and note the temperatures shown by the thermometers each minute for 5 minutes. Which flask has taken up most heat energy?

Has the heat travelled to the flasks by conduction or convection? Why not?

Remember the direction you would expect hot air to move in.

The heat has again travelled by radiation. What has the silver paint done to the heat energy? (It may help you to remember that mirrors have a layer of silver deposited on them.)

This experiment tells us that dull dark surfaces also absorb (or take in) heat energy better than do polished bright surfaces.

Experiment 9.10
The vacuum flask

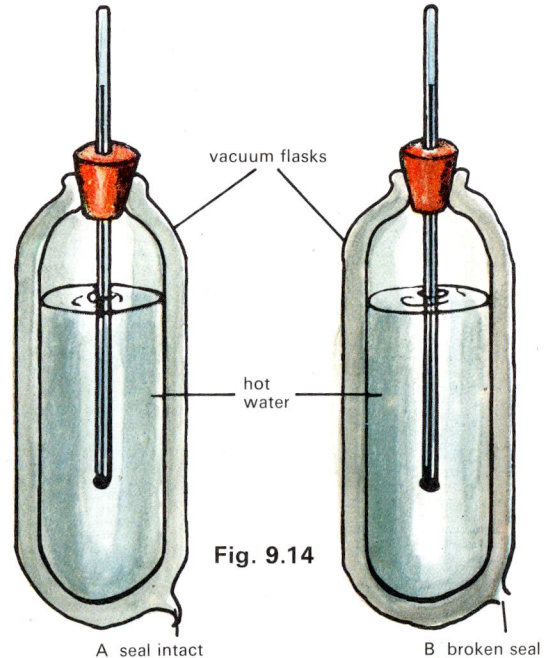

vacuum flasks

hot water

Fig. 9.14

A seal intact   B broken seal

Fill each flask with equal volumes of nearly boiling water and stopper each one as shown in Fig. 9.14. Note the temperature on each thermometer at intervals of 1 minute for 5 minutes. Which flask has lost its heat quicker?

Look at the flasks and see what is the difference between them. Because the seal is broken, air has entered the space between the glass walls in one of the flasks. The other one is perfect; there is a vacuum between the glass walls. How may heat from the hot inner glass wall have travelled to the outer glass wall in the case of the broken flask? Can you detect any difference of temperature between the outsides of the flasks by touch?

In flask A, radiation is prevented by the silver walls, while the vacuum prevents loss or gain of heat by conduction or convection.

Fig. 9.13 Why do you think houses in hot climates, like these in Tunisia in North Africa, are painted white?

Do you know another use for a vacuum flask besides for keeping things hot?

Find out from a book in the library for what purpose the Scottish scientist, Sir James Dewar, invented this kind of flask.

Experiment 9.11
A peculiar light bulb

In this experiment you will use an electric light bulb, one half of which has been coated with black paint or paper (Fig. 9.15). Place the

Fig. 9.15

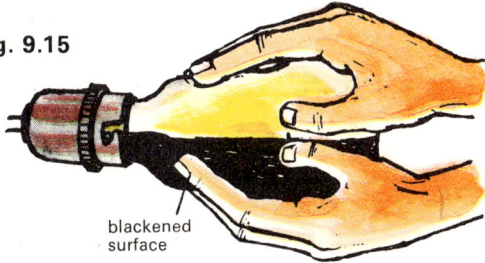

blackened surface

palm of each hand one on each surface of the bulb. Do they feel any different? Now ask a partner to switch on for about 5 seconds. Do both hands still feel the same?

Which surface is giving away more energy? How does this experiment remind you of Experiment 9.9?

Experiment 9.12

Look at the diagram in Fig. 9.16. Stick two drawing pins with candle wax or vaseline about 15 cm up the backs of the metal sheets, A and B.

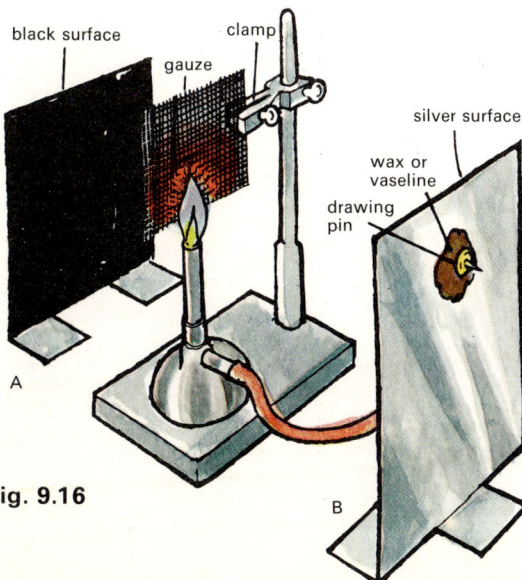

black surface

clamp

gauze

silver surface

wax or vaseline

drawing pin

A

Fig. 9.16

B

Use a Bunsen with a blue flame to heat the clamped gauze, or use an electric heating element like that used in Experiment 9.3. Put the metal sheets 15 cm on either side of the heat source. Which pin falls off first?

Which sheet has taken in the more heat? Compare the result of this experiment with that of Experiment 9.9.

Experiment 9.13

Here we use a cylinder which is wooden at one end and metal at the other (Fig. 9.17). Wind a sheet of paper once round the cylinder.

Warm the cylinder in a yellow Bunsen flame for a few seconds until you find the paper beginning to char. Against which part of the cylinder has the paper charred most? Touch the parts of the cylinder where you applied the heat. Which feels the hotter? Where has the heat gone to that you supplied to the paper next to the metal? Has the wood conducted away any of the heat given to the paper round it?

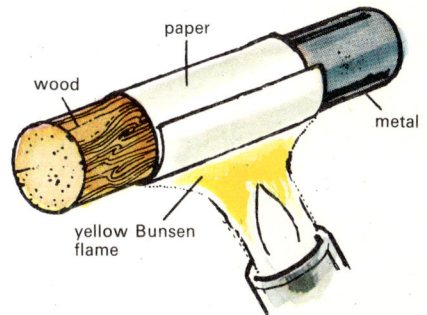

paper

wood

metal

Fig. 9.17

yellow Bunsen flame

The paper round the metal part of the cylinder does not burn easily because the metal conducts the heat away from it.

Try putting one hand on the wooden bench top, or a stool, and the other on a gas tap or a water pipe. Does one feel colder than the other? As both are in the same room and have been for a long time, it is very likely that they will both be at the same temperature; yet they do not feel like it. Why does the metal feel colder? Remember that when we feel cold this means that heat is being led away from the body.

9.4 HOW ANIMALS KEEP WARM

In Unit 3 we learned to identify the different forms of energy that could exist. From what do we obtain our own energy? In Unit 8 we found that much of this energy from food is transformed in our bodies into heat. We are now going to find out whether the coverings which different animals have make any difference to the rate at which they lose heat.

Some animals are covered with hair, others with fur or feathers. As it may be difficult to get hair for this experiment (unless you are friendly with a barber) we shall use cotton wool instead – but if you can get hair so much the better.

Fig. 9.18 How are these animals insulated against the cold?

Experiment 9.14
Hair, fur, or feathers?

We need four large flasks (500 cm³). Wrap some cotton wool (or hair) round one of them, fur round another, and feathers round the third. Leave the fourth one uncovered (Fig. 9.19). Pour 500 cm³ of nearly boiling water

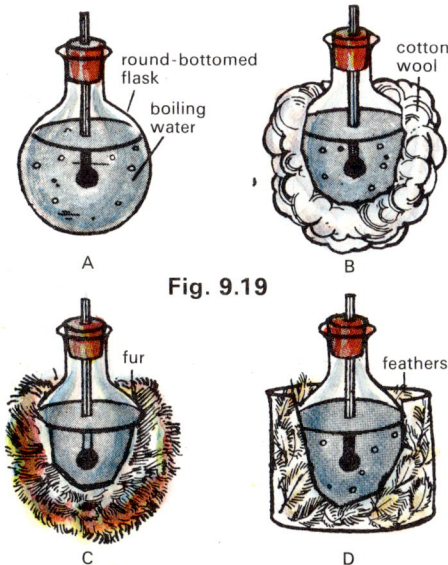

Fig. 9.19

into each flask, and fit the stoppers and thermometers. Make sure that each thermometer bulb is in the water.

Write down the temperature of each flask at intervals of 1 minute for 15 minutes. You can plot these temperatures as four graphs if you like, using a different coloured pencil for each one.

Time interval (minutes)	Flask A	Flask B	Flask C	Flask D
Start				
1				
2				
3				
4				
5				
6				
7				
8				
9				
10				
11				
12				
13				
14				
15				

Which flask cooled most in the 15 minutes — and therefore which one lost most heat? Which material made the best insulator?

Experiment 9.15
Which is the best for keeping things hot?

We are going to use different containers (Fig. 9.20) and see which is best for keeping things hot. Our containers will be a metal can, an

Fig. 9.20

'aerocup' which is often used as a cup at picnics, and is made of expanded polystyrene, a 'lagged' metal can, and an aerocup with a lid.

Put equal volumes of nearly boiling water into the various containers, and write down the temperature of the water in each container at intervals of 1 minute for a period of 10 minutes. Record your findings in a table like the one you used in Experiment 9.14. Which container has lost most heat? Which one has retained heat best? Why?

Comparing the results from different containers in pairs, try to explain why the following have not lost equal amounts of heat: A and C, A and B, B and D.

9.5 SOME QUESTIONS FOR YOU

If you have really grasped the work of this unit you should be able to answer these questions – so test yourself.

1. In which would water boil faster, an iron or a copper pan?
2. With what would you lag a water pipe in winter? Why?
3. Which would you expect to be the hotter, air near the floor, or near the ceiling?
4. Look up Unit 4 and find out what the Montgolfier brothers did, and how it worked.
5. Find out about cavity walls and double glazing. What is the material both make use of to prevent heat transmission?
6. Why are
   (a) teapots often shiny and polished,
   (b) the backs of refrigerators and motor cycle engines usually painted black,
   (c) tennis clothes white?
7. Say which of conduction, convection, or radiation is involved in
   (a) taking a hot bath,
   (b) ventilating a room,
   (c) using an 'infra-red' lamp,
   (d) using a soldering iron,
   (e) using a tea cosy,
   (f) using a fireman's asbestos suit,
   (g) using a spaceman's suit.
8. Why do birds look 'fat' on a cold day?
9. Look at Figs. 9.21 and 9.22 and answer the questions underneath them.

WHAT YOU HAVE LEARNT IN THIS UNIT

1. Heat can travel in three ways:

(a) **Conduction:** particles in the heated part of the material vibrate more energetically and in turn make their neighbours vibrate more. In this way the energy is passed from particle to particle. Metals are the best conductors. Liquids and gases are poor conductors. Many insulators contain air.

(b) **Convection:** heat energy is carried by currents of particles, and so convection can only occur in liquids and gases (fluids). The hot, expanded fluid has a smaller density than the cold fluid, and flows upwards as the cold fluid sinks.

(c) **Radiation:** the heat energy travels in invisible waves and travels from hot to cold regions. It needs no material to carry it from one point to another, and can therefore travel through a vacuum. This is how heat reaches us from the sun.

2. Heat is conducted quickly in copper, slower in aluminium, brass, and iron in that order.

3. Dull dark surfaces both radiate and absorb heat better than bright polished surfaces.

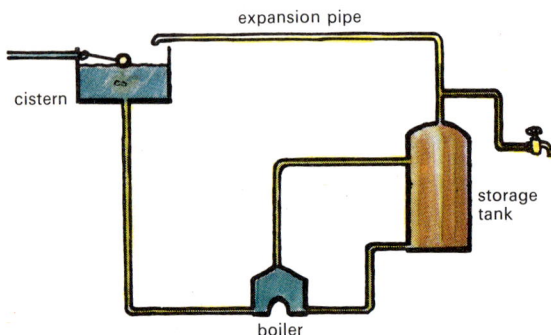

Domestic Hot Water System

Fig. 9.21

Find out the direction in which the water flows in each pipe.

sea breeze during daytime

Fig. 9.22

Can you puzzle out in which direction the breeze blows at night?

Unit Ten
Hydrogen, Acids, and Alkalis

10.1 ANOTHER GAS

In this unit we are going to study a gas that you have met before, but about which you as yet know very little. It is **hydrogen** – a very interesting and useful gas. Do you remember where you came across it before?

When you hear the word 'hydrogen' perhaps you think straight away of hydrogen bombs. Hydrogen bombs do not use ordinary hydrogen but a special form of it which is very difficult to make. However, in spite of this, you will have a lot of fun with hydrogen because it is a gas with which we can make pops, bangs, and squeals, as you will soon find out.

10.2 SOME FACTS ABOUT HYDROGEN

First of all let us find out a few things about hydrogen. Does it dissolve in water? You should be able to find this out for yourself.

Experiment 10.1
Does hydrogen dissolve in water?

Your teacher will provide you with a test-tube full of hydrogen and a basin of water. Find out if the gas dissolves.

Your test-tube of hydrogen may have been filled from a cylinder of the gas, or from a large glass bottle full of the gas, which we call an aspirator. Remembering what you have discovered in the last experiment, how, do you think, you could fill a test-tube with hydrogen? Another thing we would like to know is whether the gas will burn or not.

Experiment 10.2
Does hydrogen burn?

Hold a lighted taper near the mouth of a test-tube full of hydrogen. What happens? Now get another test-tube which is almost full of hydrogen. Hold the tube with its mouth downwards and remove the cover over the end of the tube so that the water runs out. Then hold a lighted taper near it. What happens this time?

So hydrogen burns. Your teacher will show you that things will not burn in it. This is best done with a gas jar of hydrogen rather than a test-tube. If a lighted taper is pushed into a jar of hydrogen held upside down, the gas burns at the mouth of the jar, but what happens to the taper?

Compare this with what happens with oxygen. They seem to be exact opposites as far as burning and letting things burn are concerned.

You found out that when a mixture of hydrogen and air is burnt it made a peculiar squealing noise. This is a slight explosion. If you get the proportion of air just right you can get a very violent explosion. You know already what gases are present in the air, and you know what burning really is. Which gas in air combines with hydrogen when it burns? To get the loudest explosion would you use air to mix with the hydrogen, or only part of the air? Which part?

A good idea of the force of the explosion can be got by filling a plastic bag with a mixture of hydrogen and oxygen – two-thirds of hydrogen and one-third of oxygen – and putting a light to it. This is a dangerous experiment, and you should not try it yourself. If your teacher decides to do it, it is best done outdoors so that you can stand well back. You will not blow up the school, but you will have all the other classes wondering if an aircraft has just broken through the sound barrier.

When the hydrogen burns there is, of course, a flame. Perhaps you have seen it. It is a very pale blue flame and is sometimes difficult to see. It is also a very hot flame. So when hydrogen burns a lot of heat energy is given out, as well as some sound energy. We have not yet found out what the product of the burning of hydrogen is, but we can say something about the energy it contains compared with that of the elements it was made from. What can we say?

10.3 A TEST FOR HYDROGEN

You have found that when hydrogen burns it makes a peculiar noise. This is used as a test for the gas.

What must we be sure about before we can say that this is a satisfactory test? What could you ask your teacher to do if you are not satisfied?

10.4 MORE ABOUT HYDROGEN

Experiment 10.3

Take two test-tubes of hydrogen. Hold one with its mouth upwards and the other with its mouth downwards. After about 20 seconds, put a lighted taper to the mouth of each. What happens? What can you conclude about the density of hydrogen?

Hydrogen is, in fact, the 'lightest' gas known. Can you express this more scientifically? Start by saying 'Of all gases, hydrogen has . . . .', and put in three words in place of the dots.

How would you fill a dry test-tube with hydrogen, and keep it dry?

10.5 WHAT IS FORMED WHEN HYDROGEN BURNS?

Experiment 10.4

Fill a dry test-tube with hydrogen and burn it.

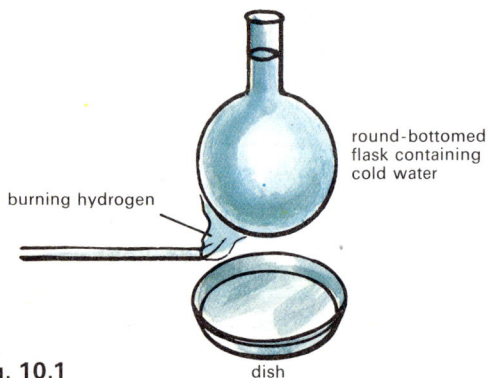

round-bottomed flask containing cold water

burning hydrogen

Fig. 10.1   dish

What do you notice about the sides of the tube? Before you lit the gas the tube was dry. Is it dry now?

What can this be? To find out we shall have to collect more of the substance, as in the experiment you did the quantity formed was very small. Your teacher will therefore burn some hydrogen at a jet, and collect some of the liquid formed. The jet of burning hydrogen must be made to play against a cold surface if we are going to collect some of the liquid.

To find out what it is we shall have to make some guesses and then see if we are right. What liquids do you know of? Write down as many as you can think of in your note book. You will probably have down things like turpentine, petrol, milk, lemonade, paraffin, water, some acids, and alcohol. You can go through your list and eliminate some of the suspects straight away because they smell, and the liquid you have collected does not. Then, some of them burn, and your liquid does not. (How do you know it does not?) So perhaps, when you have carried out a few further tests, you will find that the only liquid left is water. But this way of solving a problem – by eliminating the suspects – is not a fool-proof one, as there might always be some suspects which escape the net; there were probably lots of liquids that you forgot to put into your list, or perhaps that you had not even heard of. What we want is a positive test which will tell us that the liquid is water, and cannot possibly be anything else.

What properties does water have that no other liquid has? Do you remember the temperature at which it boils and freezes? Would these do to distinguish water from any other liquid? What about its density? It is very unlikely that a liquid will have all three of these properties the same as water, and yet not be water. If therefore the liquid we have collected freezes at 0 °C, boils at 100 °C, and has a density of 1 g cm^{-3}, it is very likely indeed to be water. To carry out these experiments requires a fair amount of liquid, and it may not be possible for your teacher to get enough to try all three tests; but when these tests are tried it is found that the liquid formed when hydrogen and oxygen combine is water.

Think how different water is from hydrogen and oxygen. Write out a list of the differences in your note book. Something very drastic must have happened to the hydrogen and oxygen when they combined. Remember too that a lot of energy was given out when water was formed. Perhaps these two facts are related.

10.6 WATER IN CRYSTALS

Experiment 10.5
Where does the water come from?

Take some blue crystals of copper(II) sulphate and feel them. Are they damp? Now heat some of them in a small test-tube. What happens?

If your teacher collects enough of the liquid which is produced in this experiment you can try the melting point and boiling point tests on it. If you do, you will find that the liquid is water. This is very odd, because the crystals did not feel damp. Where did the water come from? The only answer we can give is that the water must have been combined with the other things that go to make up copper(II) sulphate; and we know that when things combine together they become completely changed. Water which is combined in

crystals is called **water of crystallization**. When the substance was heated the water of crystallization was given off and the substance left is said to be **anhydrous** – a word which simply means 'without water'. The word 'dehydrated' would have done as well – and perhaps you are more familiar with this, as we often talk about dehydrated foods, dehydrated milk, and so on.

The white substance left in the tube in the above experiment is anhydrous copper(II) sulphate. What do you think will happen if you add water to it? Try it and see. Did you notice everything that happened? Did it get warm? What do you think this indicated? Does it agree with what we said about water being *combined* in the copper sulphate crystals?

If you add the liquid you get by burning hydrogen in oxygen (or air) to anhydrous copper sulphate, the latter turns blue. Is this a good enough test to show that the liquid is water? Think very carefully before you answer that question.

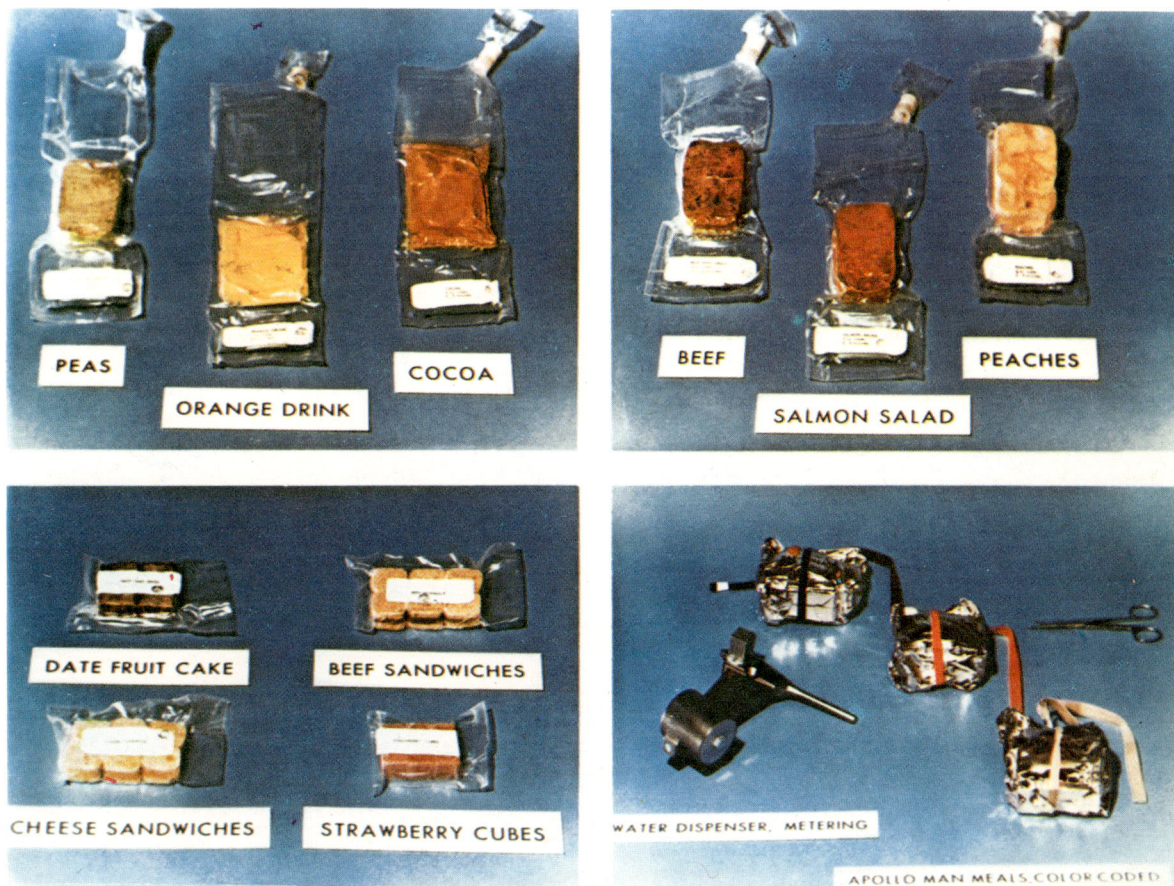

Fig. 10.2 Dehydrated foods prepared for America's astronauts

Experiment 10.6
Do all crystals contain water?

This question can soon be answered by seeing what happens when you heat different crystals. Try heating some of the following: sodium carbonate, sodium chloride (common salt), magnesium sulphate (Epsom salts), potassium nitrate, potassium chloride, sodium sulphate, potassium sulphate, cobalt chloride. You need not heat them all; different members of the class can heat different ones and you can pool the results.

10.7 SYNTHESIS AND ANALYSIS

Water is made by burning hydrogen in oxygen. A process like this where a substance is built up from simpler things is called **synthesis**, a word which means bringing together. Do you remember the term **photosynthesis**?

You will be thinking, though, that if water can be synthesized from hydrogen and oxygen it ought to be possible to break it up again into the elements from which it is made. This kind of process – breaking up – is called **analysis**.

You will remember that in Unit 4 we synthesized copper(II) chloride from copper and chlorine, and then we analysed the copper(II) chloride. When the substance was synthesized a lot of energy was given out, and when it was analysed energy had to be put back. The position is exactly the same with water. A good deal of energy is given out when water is synthesized. If we want to break it up again we shall have to put energy back.

How can we do this? By heating it? What happens to water when it is heated? Does it break up into hydrogen and oxygen? Perhaps if we heated steam very strongly it would break up. If we try this experiment in the laboratory we find that we cannot reach a sufficiently high temperature with any method of heating available; however, scientists have partly decomposed water into hydrogen and oxygen by heating it very strongly indeed.

Do you remember how we broke up copper(II) chloride? Suppose we try the same method with water.

Experiment 10.7
Breaking up water

Connect up a circuit like the one shown in Fig. 10.3. Is there any reading on the meter?

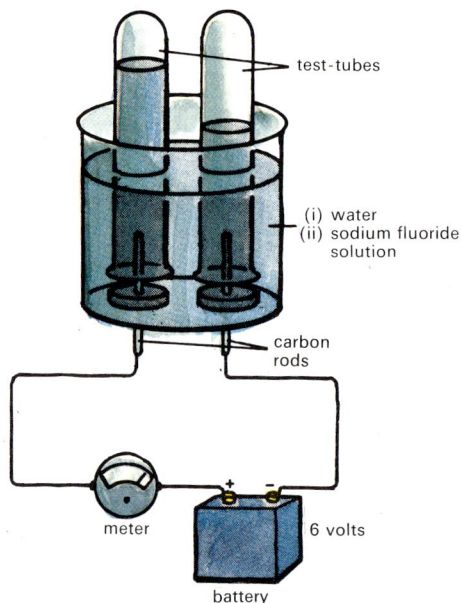

Fig. 10.3

Now add some sodium fluoride to the water. Does the meter read now? Collect and test the gases given off. Notice whether there is more of one gas than of the other and what the approximate volume relationship is. Does this agree with what you expected?

Answer the following questions: Before you conclude from this experiment what water is made up of, what must you know about sodium fluoride? To break up the water you have had to put in energy. In what form was this energy supplied?

10.8 SOME PECULIAR METALS

Some time ago you came across the metal calcium (Unit 1) and tried putting a piece of it into water. We shall repeat this experiment just to remind ourselves of what happened.

Experiment 10.8
Calcium and water

Take about 5 cm depth of water in a test-tube and add one or two small pieces of calcium which have been scraped with a knife. What happens? What did you scrape off the calcium? Why did you do this? Was energy given out or taken in in this reaction?

The white powder remaining in the tube is calcium hydroxide. You have met this substance before. Is it completely insoluble in water?

Filter off the powder, and test the filtrate with indicator paper. What does this show? Show that the filtrate reacts with carbon dioxide. How will you do this without making carbon dioxide specially?

There are some other peculiar metals which do much the same thing as calcium. You will be shown some **sodium**. Why do you think it is kept under oil? (Recall the scraping of the calcium in the last experiment.) Sodium is so soft that it can be cut with a knife.

Experiment 10.9
Sodium and water

Your teacher will cut a small piece of sodium and put it on to water in a trough. What happens? Is a gas given off? Unfortunately it is too dangerous to try to collect anything that comes off, but your teacher will try to burn it.

Test the liquid left after the experiment with indicator paper. Is the effect the same as with the calcium experiment? If so, what do you think is produced? Why is no white powder formed?

You have come across this substance before, and you have been warned that it is dangerous. However, the solution is so dilute that no damage will result if you put your forefinger and thumb into the liquid and 'feel' it. What do you notice?

In this experiment was energy taken in or given out? What kind of energy? Was the reaction more or less vigorous than that with calcium?

Experiment 10.10
Magnesium and water

Try putting some magnesium powder contained in a twist of tissue paper under water in a test-tube as shown in the diagram. Try the same thing with magnesium ribbon.

You will have to wait for some time before anything happens. In which tube does gas collect first? Can you explain this on the kinetic theory of matter? What gas comes off? Can you suggest a way of making the reaction proceed faster? What about trying boiling water?

Heat some magnesium powder with water in a test-tube. Is the reaction any faster?

Perhaps it will go faster still with steam. This can be tried with the apparatus shown in Fig. 10.4. The metal (which must be magnesium ribbon, NOT powder) is heated and the heat

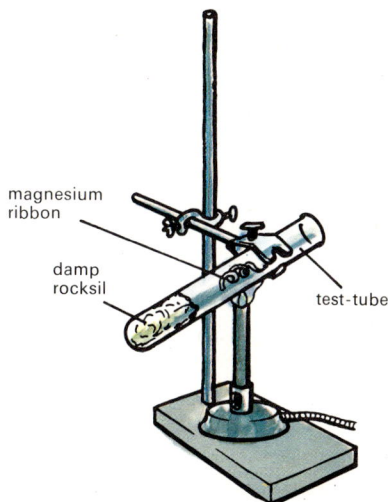

Fig. 10.4

from this section of the tube is enough to boil the water from the damp rocksil. The steam comes over the heated magnesium. What happens?

Can you see why you used magnesium ribbon and not powder? Explain in terms of the kinetic theory why the reaction with steam is so much more vigorous than with cold water. Considering the action of the metals calcium, sodium, and magnesium on *cold* water, place the reactions in order of activity, putting the most active first.

Experiment 10.11
Other metals and water

You can use the apparatus you used with magnesium to investigate the action of other metals on steam. Different groups in the class can try zinc, iron, tin, lead, and copper. Pool your results. Then see if you can agree on an order of activity for the reactions of these metals with water. You can include the three you have used before in your list.

You will have found that some of these metals will not decompose steam at all, so we get an order of activity like this:

> sodium
> calcium
> magnesium
> zinc
> iron
> _____
> tin, lead, copper — no result

Why can we not put the last three into order?

10.9 A CHEMICAL TUG-OF-WAR

When a metal decomposes water or steam it turns out the hydrogen, as we have seen. What happens to the oxygen? Look at the residue in the tube when you passed steam over heated magnesium. Have you seen anything like it, connected with magnesium, before? So

$$\text{magnesium} + \text{water} \rightarrow \text{hydrogen} + \text{magnesium oxide}$$

In the case of sodium and calcium we said that the hydroxides were formed. This is because the oxides of these metals combine further with water to form hydroxides.

We see, then, that these reactions represent a competition between hydrogen and the metal for oxygen. Some metals can pull oxygen away from its combination with hydrogen in water, and some are not powerful enough to do this. It is as if Tom was Mary's boy friend. They are walking down the street together when Mary sees another boy, Robert, whom she likes better than Tom, so she says good-bye to Tom and walks off with Robert. It is very hard luck on Tom. Similarly, hydrogen and oxygen are going along together arm-in-arm. Along comes magnesium for which oxygen has a greater liking than she has for hydrogen, so off she goes with it, leaving the hydrogen alone.

Make a list of those metals you know which *can* turn out hydrogen from water and those which cannot.

One of the metals which cannot turn out hydrogen you have found to be copper. What would you expect to happen, then, if you passed hydrogen over heated copper(II) oxide? Devise an apparatus for seeing if you are right.

(1) Tom and Mary walking down street. Tom looks happy. Robert comes towards them.

Fig. 10.5

(2) Mary walks off with Robert and leaves Tom behind. Robert now looks happy and Tom disconsolate.

Experiment 10.12
Hydrogen vs. copper

You are expecting the hydrogen to take the oxygen away from the copper(II) oxide and form water. If you are to prove this, what must you be sure about regarding the hydrogen and the copper(II) oxide?

To make sure that the copper(II) oxide is dry, your teacher will have heated some in a dish and allowed it to cool in a dry atmosphere. This is done in a vessel called a desiccator. To dry hydrogen it is passed through a drying agent called silica gel. You have met this before in Unit 5.

10.10 ACIDS

Acids have often been mentioned in this book. Make a list of any you can remember. Have you included sulphuric, hydrochloric, and nitric acids? These are the commonest acids we use in the laboratory and we have met all of them – one as recently as the last paragraph.

The word 'acid' means 'sour'. We often say that a sour liquid tastes 'acid'. Are there any sour liquids you come across at home? What about vinegar, lemon juice, grape-fruit juice, tartaric acid, citric acid, and sour milk? All these are acids, or mixtures of acids. Vinegar contains acetic acid, lemon juice and grape-fruit juice contain citric acid (you have heard of citrus fruits), and sour milk contains lactic acid.

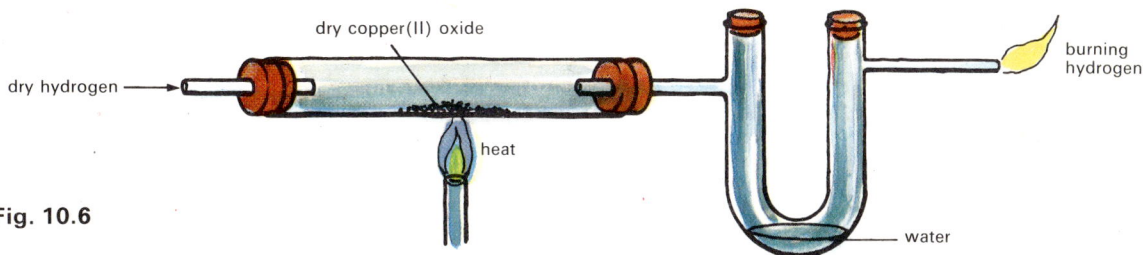

Fig. 10.6

The apparatus is shown in Fig. 10.6. Is it anything like the one you thought of? When the hydrogen has swept all the air out of the apparatus the copper(II) oxide can be heated, but not before. Why is this? What happens in the U-tube? You think water may have collected. What tests would you carry out on the liquid to find out?

What has happened to the copper(II) oxide? Some time ago (Unit 4) you saw what happened when nitric acid was added to copper, but if you have forgotten, try it again. In one test-tube put some copper turnings, in another some copper(II) oxide, and add some dilute nitric acid to each. Can you use the results of this experiment to distinguish between copper and copper(II) oxide? When the apparatus is cool, take out the brown powder and try the nitric acid test on it.

Sum up in one line what has happened when hydrogen was passed over heated copper(II) oxide. What would you expect to happen if you passed hydrogen over heated magnesium oxide, and over heated lead oxide?

Fig. 10.7 A nitric acid plant

Experiment 10.13
Testing for acids

Although we have said that acids taste sour, it is not very wise to use the sense of taste to find out if a substance is an acid because some acids are very strong and can burn you badly, and some are poisonous. We never taste things in the laboratory unless we are quite sure that they are harmless. An indicator paper which is dipped into the liquid is used instead.

Collect samples of as many of the acids mentioned above as you can and dip a piece of indicator paper into each. Say what happens to the paper. Now try putting the paper into water and a solution of salt. Can the paper be used to tell whether a liquid is acidic?

We use a scale to measure acidity. It is called the pH scale. If a solution is acidic the pH is less than 7. Your indicator paper shows different colours at different pH values. Look at the chart on the book of papers. The acid solutions you have tested have all been strong and have produced the same shade of colour with the indicator.

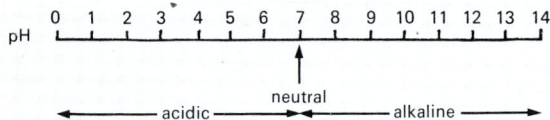

Fig. 10.8

Experiment 10.14
Testing a solution of carbon dioxide in water

You will be provided with a solution made by bubbling some carbon dioxide into water. Test it with indicator paper. Is it acidic? If so, what is its pH value? You can test some lemonade in the same way.

Experiment 10.15
Does water play a part in making substances acidic?

All the substances you have tested so far have contained water. It seems fair to ask if the water has anything to do with making them acidic. Try touching some citric acid crystals, some tartaric acid crystals, and some oxalic acid crystals with dry indicator paper. What is the result? Now try touching the substances with a moist indicator paper. What happens now? What is the answer to our question?

You will come across the reason for this effect later on in the chemistry course.

10.11 METALS AND ACIDS

All acids contain hydrogen, so perhaps we can obtain hydrogen from them. Can you suggest how this might be done? Here is a clue: how did you get hydrogen from water?

As a try-out let us see if magnesium will release hydrogen from an acid. We could take almost any acid, but dilute hydrochloric acid is very convenient for the job.

Experiment 10.16
Will magnesium turn out hydrogen from dilute hydrochloric acid?

You can work out how to do this for yourself. What happened? Compare the rate of this reaction with that of magnesium on water. Would it be safe to try the action of sodium on dilute hydrochloric acid? What about other metals? Will they turn out hydrogen from dilute hydrochloric acid?

Experiment 10.17
The action of metals on dilute hydrochloric acid

Different members of the class should try aluminium foil, iron filings, lead shot (or foil), tin foil, copper foil, silver foil, and a drop of mercury. Remembering what happened with steam, are the results what you expect? Write your results in a table like the one on page 19. You have confirmed the activity series that you arrived at before with another reaction. If you think back to the burning of metals in oxygen you will remember that a similar order was obtained for the vigour with which the metals reacted with oxygen. Magnesium burnt most

Metal	Reaction (yes/no)	Rate (fast/slow/nil)

Fig. 10.9

vigorously, while copper did not burn at all (although it turned black). Would you expect a tie-up like this? Explain why you would.

Experiment 10.18
Metals and dilute sulphuric acid

If you have time, find out whether dilute sulphuric acid reacts with metals in the same way as dilute hydrochloric acid. Do you get the same order of activity?

If you have been thinking like a scientist, you will have been curious to know what happened to the magnesium when it reacted with the dilute hydrochloric acid. There is obviously no magnesium in hydrogen. (How do we know? You might answer 'Because hydrogen is an element'. But how do you know that? A more satisfactory answer can be obtained by thinking of the action of dilute hydrochloric acid on other metals besides magnesium.) Does it look as though there is magnesium in the liquid remaining? Can you see any? Perhaps it is in solution, like salt in water. How can we find out?

Experiment 10.19
What happens to the metal when it reacts with an acid?

Pour some of the solution remaining in the test-tube on to a clock glass and stand the latter on top of a beaker or can containing water (Fig. 10.9). Boil the water. The steam heats the liquid and causes the water to evaporate. Is there a residue? Does it look like magnesium?

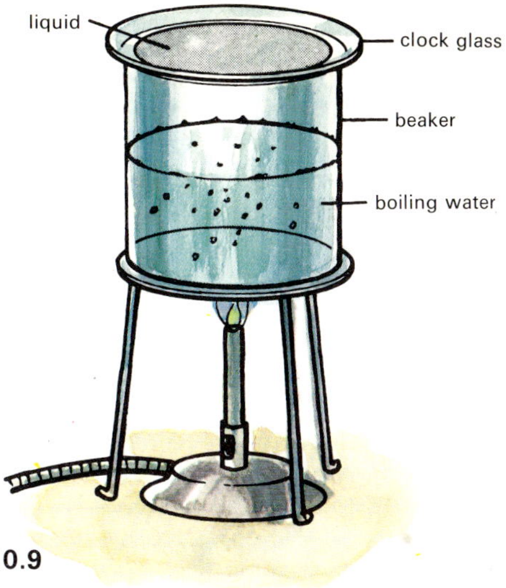

Quite clearly something drastic has happened to the magnesium. As it has not escaped from the vessel it must still be there in some form or other. Yet it is not there as the metal magnesium. You will have found that a white substance is left. It must clearly contain the magnesium combined with something else. It is, in fact, magnesium chloride.

We can understand how it has been formed if we remember that hydrochloric acid is a solution of a substance called hydrogen chloride in water. The magnesium has turned out the hydrogen from the hydrogen chloride and has combined with the chloride part of the substance. We can put it like this:

magnesium + hydrogen → magnesium + hydrogen
        chloride   chloride

Here we have competition coming in again. The magnesium has a stronger pull on the 'chloride' than hydrogen has, so it changes places with it.

The magnesium chloride is called a **salt**. There are many different salts; the most common of them is sodium chloride, which is called common salt.

Dilute sulphuric acid is a solution of hydrogen sulphate in water. What salt do you think is formed when magnesium is acted upon by dilute sulphuric acid?

Magnesium sulphate is sold in shops as Epsom salts. Find out what it is used for.

10.12 ALKALIS

Experiment 10.20
More tests

Here is a list of substances you will find in the laboratory: solutions of sodium hydroxide, sodium carbonate, lime water, sodium bicarbonate, ammonia. Try the effect of them on indicator paper. Are they acids?

You will find that these solutions have a pH greater than 7, whereas acids have a pH less than 7. They are called **alkalis**. They are in some ways the opposite of acids. What do you think will happen if you mix an acidic solution with an alkaline one? If they are opposites perhaps they will cancel each other out. Let us find out what does happen.

Experiment 10.21
What happens when acid is mixed with alkali?

You will be provided with a dilute solution of sodium hydroxide and some dilute hydrochloric acid. With the universal indicator find the pH of each.

Now take 10 cm³ of the dilute sodium hydroxide solution in a test-tube, and add a drop or two of universal indicator. Fill a graduated syringe with the acid solution and add it 1 cm³ at a time to the sodium hydroxide solution. Notice the colour of the indicator after each addition. Put down your results in a table like this:

Volume of acid added (cm³)	Colour	pH
0		
1		
2		
3		
4		
.		
.		
.		
.		
.		

Does the acid 'cancel out' the alkali? How much did you have to add for this to happen? What happens if you add more acid than this? We call this 'cancelling out' **neutralization**. The acid has neutralized the alkali — and, of course, at the same time the alkali has neutralized the acid.

Was a definite amount of acid required to neutralize the 10 cm³ of alkali? Did all members of the class get the same result?

10.13 WHAT IS PRODUCED WHEN AN ACID NEUTRALIZES AN ALKALI?

As both the acid and the alkali are destroyed, something new must be formed.

Experiment 10.22

Take 10 cm³ of the same sodium hydroxide solution you used for the last experiment and add to it the volume of hydrochloric acid that you found just neutralized it, but this time do not add any indicator. Stir, and find the pH of the solution by taking out a drop on a glass rod and touching it on indicator paper. If you have worked accurately it should be neutral, but if it is not add a drop or two of alkali or acid, whichever is required to make it really neutral. Taste a drop of the solution. What is it like? To see if you are right, make a weak solution of sodium chloride, taste it, and find its pH.

Is this a good test for showing that the product of neutralization of sodium hydroxide by hydrochloric acid is sodium chloride? Why not? There are better ways of showing that this is what is now in the solution; you will meet them later.

In this experiment you have been asked to taste something. As has been said frequently before you should never taste anything in the laboratory unless you have been told specifically that it is safe to do so. In this case it is perfectly safe.

How do you think you would make sodium sulphate, calcium chloride, and potassium nitrate?

In Experiment 10.21 you found that all members of the class obtained the same result for the volume of acid required to neutralize a certain volume of sodium hydroxide solution. This was only to be expected as you were all using the same solutions. What would have happened if some members of the class were using solutions of different concentrations?

Suppose your group was using 10 cm³ of a solution which contained 0.5 g of sodium hydroxide, while your neighbours used 10 cm³ of a solution which contained only 0.25 g of sodium hydroxide, but both groups neutralized their solutions with hydrochloric acid of the same concentration. Do you think the second group would require more, less, or the same volume of acid to neutralize their solution? Let us see if you are right.

Experiment 10.23

You will be provided with a solution which contains 1 g of sodium hydroxide in 20 cm³ of solution. Measure out 10 cm³ of this solution with a syringe. What mass of sodium hydroxide does it contain? Find out how many cm³ of hydrochloric acid are required to neutralize it.

Now take 10 cm³ of the sodium hydroxide solution, add 10 cm³ of water to it and stir. What mass of sodium hydroxide is present in the 20 cm³ of solution?

Measure out 10 cm³ of the solution with a syringe. What mass of sodium hydroxide does this 10 cm³ contain? Find out what volume of hydrochloric acid is required to neutralize it.

Compare the masses of sodium hydroxide taken in these two experiments with the volume of acid required to neutralize them. Is there any connection? Were you right in your answer to the question above?

Is it the volume of solution that matters, or the mass of sodium hydroxide present? In the last experiment you took the same volume of sodium hydroxide solution (10 cm³ each time) but the solutions were of different concentrations. To answer our question we must take solutions of sodium hydroxide which contain the same mass of solute in different volumes of solution.

Experiment 10.24

Measure out 10 cm³ of the sodium hydroxide solution into each of three beakers, A, B, and C. To beaker B add 20 cm³ of water and to beaker C add 40 cm³ of water. Now find out what volume of hydrochloric acid is required to neutralize the alkali in each beaker.

What can you say about the mass of sodium hydroxide present in each beaker? What did you find about the volume of hydrochloric acid needed to neutralize the contents of each beaker?

We started with 10 cm³ of sodium hydroxide solution. Does the volume of water added to this solution make any difference at all to the mass of hydrochloric acid required to neutralize it?

The neutralization of an acid by an alkali is a chemical reaction. This is the first occasion that we have actually dealt with the masses of substances that react together. We have found that a given mass of sodium hydroxide always reacts with a given mass of hydrochloric acid. The same is true of any chemical reaction, although we do not have time to prove it. Thus 24 g of magnesium always combines with 16 g of oxygen; 1 g of hydrogen always combines with 8 g of oxygen; 56 g of iron always combines with 32 g of sulphur to form iron(II) sulphide; 63.5 g of copper always combines with 71 g of chlorine to form copper(II) chloride.

This is a very important chemical law, called the **law of constant composition**. Does it fit in with the particulate theory of matter?

What weight of magnesium would combine with 8 g of oxygen? What weight of oxygen would combine with 3 g of hydrogen?

10.14 APPLICATIONS OF NEUTRALIZATION

Acids are corrosive substances. If you were to drop some accumulator acid (dilute sulphuric acid) on your clothes it would soon make a hole in them. What would you do to prevent this? Would you add sodium hydroxide? If you have forgotten about this substance look back to page 15. What about adding ammonia or sodium bicarbonate?

Why are these used instead of sodium hydroxide, although all are alkalis and would neutralize the acid?

Fig. 10.10

Indigestion is often caused by the presence of too much acid in the stomach. You may have seen in some advertisements for indigestion remedies that they 'relieve acidity'. What kind of substance do you think might be present in these medicines?

The presence of acid in the soil has a considerable effect on the readiness with which plants grow in it. How could you find out the acidity of the soil?

Experiment 10.25
Testing soil for acidity

You should be able to work out how to do this. Different members of the class could bring in a little soil from their gardens for testing. Are they all equally acid?

Farmers often have to reduce the acidity of the soil to get the best results for their crops. Write in your note book which one of the following you would add:

sodium hydroxide
sodium carbonate
calcium hydroxide
 (slaked lime)

ammonia
hydrochloric acid
salt

Give a reason for your answer and explain why you did not choose the others.

Fig. 10.11 Crop spraying

WHAT YOU HAVE LEARNT IN THIS UNIT

1. Hydrogen is a gas which burns in air with a pop, and does not dissolve in water. It has a very low density.

2. Water is formed when hydrogen burns. This suggests that water is an oxide of hydrogen.

3. Some metals, such as calcium and sodium, decompose cold water, one of the products being hydrogen. Other metals, such as magnesium, aluminium, zinc, and iron, will decompose steam. Copper and silver *cannot* do this.

4. The metals can be placed in an order of reactivity according to the ease with which they break up water. This order is:

sodium, calcium, magnesium, aluminium, zinc, iron, copper, silver.

5. Hydrogen can also be obtained from dilute acids by the addition of certain metals. The usual way of getting hydrogen is by the action of zinc or magnesium on dilute hydrochloric acid.

6. The order of reactivity of metals with acids is the same as that with water.

7. Water can also be split up by electrolysis.

8. If hydrogen is passed over heated copper(II) oxide, copper and water are produced.

9. All acids contain hydrogen.

10. Acids neutralize alkalis. This process is of everyday importance, for example in neutralizing spilt acid, in indigestion remedies, and in correcting the acidity of the soil.

11. The mass of hydrochloric acid which neutralizes a given mass of sodium hydroxide is always the same. This is true of all chemical reactions. The weights of substances reacting to produce a given compound are always in the same ratio. This law is known as the law of constant composition.

Unit Eleven
Detecting the Environment

11.1 HOW DO WE FIND OUT?

In the preceding units you have been finding out about yourself and your surroundings, and you will remember that right at the beginning of our work, in Unit 1, we said that all our information comes to us through our senses. We found that occasionally our senses could mislead us, and we shall deal with this point again later in this unit. We are now going to find out some fascinating things about our senses. We will start by investigating the sense you are using while reading this book. Which one is that?

11.2 THE EYE AND LIGHT

Can you imagine what life must be like for those boys and girls who cannot see? We too often take our ability to see for granted, and do not take care of our eyes. In some science experiments with acids and glassware we may have to wear goggles for protection, as our eyes can sometimes be easily damaged. Of course, as some of you who play wild games know, the eye can stand quite a lot of rough treatment. Nevertheless it is the most sensitive instrument that the body has to find out about its environment, and we must be very careful how we look after it. To know more about your own eyes you will either see your teacher cut up a bullock's eye, or you may do this for yourself. Perhaps you will not like the idea of doing an experiment like this, but once you begin the dissection these misgivings will soon disappear.

Experiment 11.1
Dissecting a bullock's eye

Before you cut open the eye, have a close look at it, and try to recognize all the parts in Fig. 11.1. Handle the eye and notice that it feels rubbery to the touch, or as if it were made from tough plastic. In fact, the white outside layer of the eye (called the sclerotic coat) is very tough.

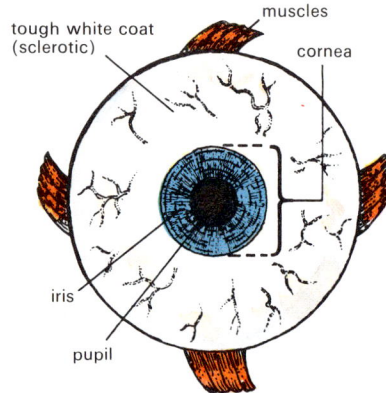

Fig. 11.1

Why? To find out about the inside of the eye we must cut through this tough coat. The tools we need must be very sharp and must be handled carefully.

Hold the eye above a glass dish and begin with a small cut with a razor blade or a very sharp scalpel just behind the iris and the clear cornea. Now continue making a circular cut round the eye. Scissors can be used just as well to do this. This circular cut will remove the clear cornea, and a watery liquid (the aqueous humour) will run out into the dish.

You will probably find that the iris is removed with the cornea, and the black rear surface should be visible with muscles lying just like the spokes of a wheel. The coloured iris can be removed from the cornea by scraping round the edge with the scalpel, and if washed in water, it can be turned over to show the coloured front side. When this is done, you can see that the black spot in the middle of the iris is simply a hole through which light enters the eye. What is this hole called?

Behind the iris you can see a little ball of jelly resting in a bed of more soft jelly, called vitreous humour (these names, aqueous humour and vitreous humour, are very old and go back

24

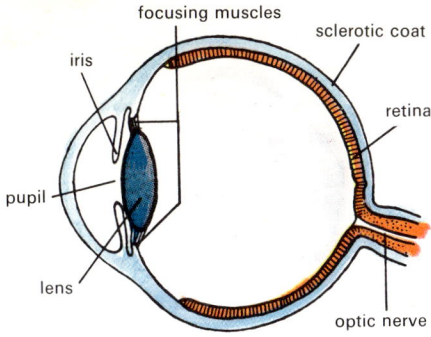

Fig. 11.2

to the time when doctors first found out what the eye was like; we still use them today). This little ball of jelly is called the lens. Pick it out with a pin or with forceps.

If you wash out the vitreous humour from the hollow eye-ball you will see the inside layer of the eye-ball, the retina. The retina is lined with very sensitive nerves.

Hold up the lens on a pin and try placing it just above a sheet of paper. Do you notice anything appearing on the paper below the lens when the classroom lights are switched on? Write your name in block capitals on the paper

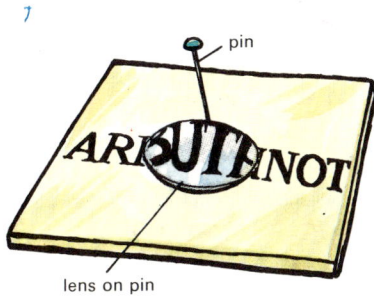

Fig. 11.3

and place the lens on top of the writing. What effect does the lens have? Do you see that it acts rather like a burning glass and brings the rays of light from a lamp together on to the paper? It also acts rather like a magnifying glass with the writing.

You found that the dark pupil in the eye is actually an opening which lets light into the eye and so through the lens and on to the retina. Look at a partner's eye when near a light and then in a dark corner of the room. What do you notice that the pupil does? The purpose of the iris is to adjust the size of the pupil. In bright light the iris closes up the pupil to let only a

little light into the eye, and so protect the retina. In dull light it opens up the pupil to allow more light on to the retina.

Returning to the bullock's eye, can you see the optic nerve which carries the message about the image on the retina to the brain?

11.3 THE PIN-HOLE CAMERA

Experiment 11.2
Making a pin-hole camera

Nowadays scientists are able to make instruments to do marvellous things, but no-one has yet been able to devise anything which works just as wonderfully as your eyes.

To try to understand how the lens works you should first of all set up the apparatus shown in Fig. 11.4. A chalk box serves the purpose very

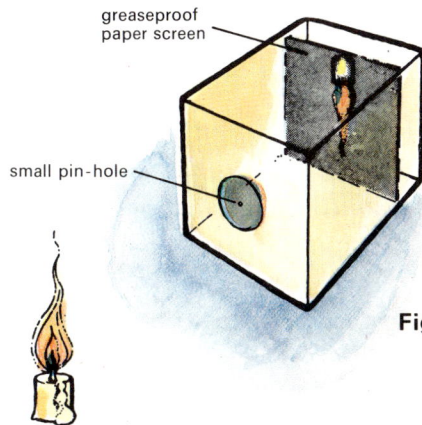

Fig. 11.4

well. Cut a square hole in one end and cover it with greaseproof paper. In the opposite end cut a round hole about the size of a penny and stick over this a piece of thick black paper held down with a piece of adhesive tape. Make a small pin-hole in the black paper and hold this towards a burning candle. It is a good idea to have the room dimmed out at this stage so that you can see more clearly. What can you see on the greaseproof paper screen?

This apparatus you have made is called a pin-hole camera. Later on you could perhaps put a piece of document printing paper at the back of the box instead of the greaseproof paper, and make a photograph of the candle flame.

You will probably see the image of the candle flame rather smaller than the flame itself if the

Fig. 11.5

box is some distance from the candle. Is it the right way up, or upside down?

Complete a sketch like the one in Fig. 11.5 to try to explain why the image should be upside down.

The image is probably rather faint, because not much light will be getting through the small pin-hole you made. Can you suggest how we might get more light through to the screen?

Some of you might try enlarging the hole. Others might try making several holes instead of just the one. Are either of these methods entirely satisfactory?

Although making the hole larger makes the image brighter, it makes it blurred. Making more holes simply results in more separate images each one overlapping the other.

11.4 ACTION OF THE LENS

Experiment 11.3

Slide a convex lens over the hole of the pin-hole camera (a convex lens is one that is thicker in the middle than at the edges), and move the lens gradually further away from the hole. You should find that at a certain distance you will get a bright, sharp, inverted (i.e. upside down) image of the candle flame focused on the screen.

This is how a lens works in a camera. We shall be looking at the camera in greater detail later on, but just now we can see that the lens allows a wide hole (or 'aperture') to be used to let a lot of light into the box, and at the same time, if it is in the right position, it makes the image sharp.

In the eye the lens is not movable – it is always kept in the same position, but when we want to focus sharply objects which are near to the eye, the muscles in the eye alter the shape of the lens.

Hold out a hand in front of you and extend your thumb. Now consciously focus on it with one eye, keeping the other eye closed. The thumb appears sharp – but what do you notice about other objects in the room? Do they not appear rather blurred? Again focus consciously on the more distant objects behind the thumb. How does the thumb appear this time?

11.5 LOOKING AT LENSES

Experiment 11.4
How a lens works

For this series of experiments you should use a ray box (Fig. 11.6). Arrange this so that three

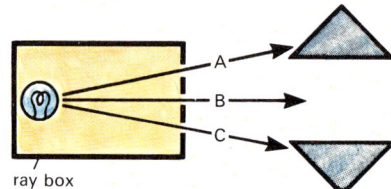

ray box   Fig. 11.6

divergent rays are sent out (divergent means that the rays are becoming more and more separated; convergent means that they are being brought closer and closer together). Now place a small prism as shown in the diagram. Does it change the direction of the ray? Keeping the prism in place, mark the ray by putting several pencil dots along it and then joining them up with a pencil and ruler.

Now put the prism at position C and do the same. What do the rays do once they have passed through the prism? Do you see that they are made to come towards each other; in other words they are made to converge?

Now place a lens like the one in Fig. 11.7 in front of the rays. This type of lens is called a

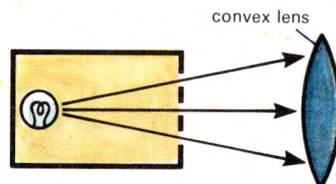

convex lens

Fig. 11.7

convex lens. Trace in the rays with a pencil when they have passed through the lens. Repeat this experiment with a thicker and a thinner lens. Which kind of lens has the greatest effect on the rays? We say the thicker lens is the more powerful. Why?

Experiment 11.5
The jelly lens

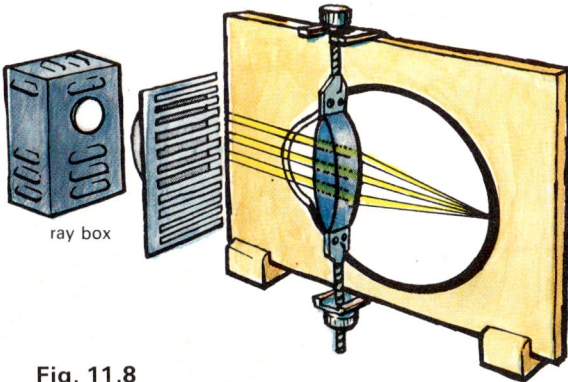

Fig. 11.8

If you have a jelly lens like the one shown in Fig. 11.8, try adjusting the screw. This affects the thickness of the lens. As you screw it up the lens gets thicker; as you unscrew it it gets thinner. See how this affects the converging light rays.

When the lens is thick the rays converge to a point more quickly. This point is called the **focus** of the lens, and the distance from the centre of the lens to this point is called its **focal length**. Does the thick lens have a shorter or a longer focal length than the thin one?

This is exactly what happens in your eye when you have to focus objects at different distances on the retina. The thickness of the lens is altered by the muscles of the eye. When you focus on near objects the lens is thicker than when you focus on distant objects.

11.6 DEFECTS IN VISION

Some boys and girls wear glasses because they cannot see clearly without them. The lens in the eye cannot focus correctly on the retina because possibly the distance between the lens and the retina is too great or too small. We can find out how to correct this by a simple experiment.

Experiment 11.6

Fig. 11.9

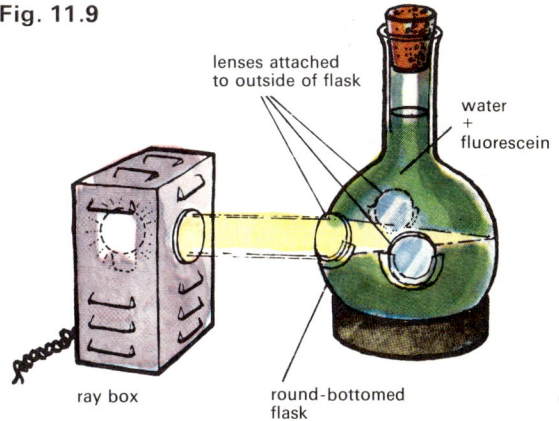

You will have a flask like that shown in Fig. 11.9 which has different lenses attached to it. Fill the flask with water to which a little fluorescein has been added. This makes it possible to see the rays passing through the water. Rotate the flask to see how each of the lenses attached to the flask affects the beam of light.

Remember the far side of the flask will represent the retina, and it is the job of the lens in the eye to focus the light on to it. Which of the three lenses is suitable for doing this?

The lens which brings the light to a focus in front of the 'retina' is showing what happens when a person is short sighted. Find a lens to put in front of the model eye to correct this and throw the point of convergence (the focus) exactly on the 'retina'. What kind of lens is it — concave or convex?

The lens which brings the light to a focus beyond the 'retina' shows what happens when a person is long sighted. Find out what kind of lens is necessary to correct this and bring the rays to a focus on the 'retina'. Is it concave or convex?

If there are boys and girls in your class who wear glasses, ask them if you can find out whether they are long or short sighted. You should be able to do this just by feeling the lenses in their glasses. A convex lens, you will remember, is thicker in the middle than at the edges, and this, you will have just discovered, is required to correct long sight. A concave lens, which is thinner in the middle than at the edges is used to correct short sight.

11.7 THE FOCUSING CAMERA

Experiment 11.7
How does a camera work?

Darken the room leaving only one window without the blinds down. Now open up a focusing camera which has no film in it. Put a piece of greaseproof paper at the back of the camera where the film would be; it can be held in position by a piece of adhesive tape (Fig. 11.10). Open the shutter and point the lens at the window, and turn the camera lens until the window frame is sharply focused on the paper screen. You can put a cloth or jacket over your head so that you can see the image more clearly. Now turn the lens so that the objects outside the room are in sharp focus. Is the frame still sharp on the screen?

Fig. 11.10

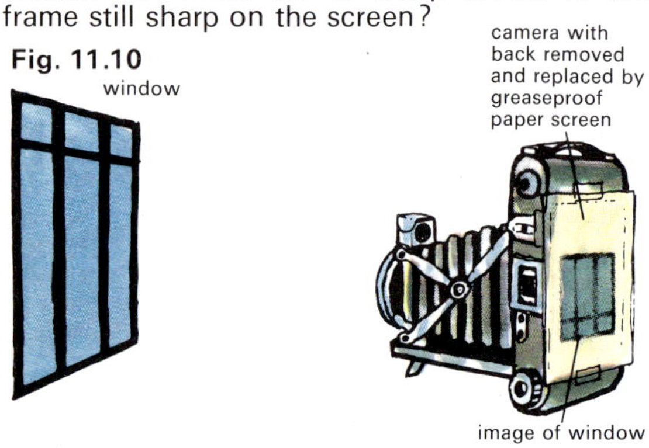

You will find that to focus distant objects you will have to screw the lens in, and to focus near objects you will have to screw the lens out.

11.8 THE RETINA

It has been found that there are different kinds of nerve endings on the retina. One of the types are called 'rods' because of their shape. They are sensitive to white light or to the absence of light (i.e. blackness). Others are called 'cones', again because of their shape. They are triggered off by lights of different colours.

11.9 COLOUR BLINDNESS

In about one in twelve males the cones do not all function properly, and such people have difficulty in distinguishing between red and green coloured things.

People who are colour blind should not worry as there must be about two million of them in Great Britain alone. Such people, however, cannot do certain kinds of job in which it is necessary

Fig. 11.11 What can you see in this picture? Are you colour blind to certain colours?

to distinguish between colours, such as aircraft pilot, railway engine-driver, ships officer, electrician, and the like.

Experiment 11.8
Are you colour blind?

You should each test yourselves with special colour blindness test-cards. How many boys in the class are colour blind? How many girls? Should you find that you are colour blind, remember to tell your careers master because it will make a difference to the kind of job you can do.

11.10 VISION. CAN YOU JUDGE DISTANCES WITH ONE EYE?

Experiment 11.9

In this simple experiment you should work with a partner. Your partner should stand facing towards you, and stretching out a hand with all the fingers clenched except the first finger, which should be pointing straight up. He should be about four or five paces away from you.

Fig. 11.12

Fig. 11.13

Stretch out a hand towards him, with all your fingers clenched except the first finger which should be pointing straight downwards. Close one eye and walk slowly towards your partner and try to place your downward-pointing finger directly on top of your partner's upward-pointing one. Do you find this easy to judge? You will very likely either stab your finger down in front of your partner's or behind it. Using only one eye it is very difficult to judge distance accurately.

Now go back to your first position in the room and repeat the experiment keeping both eyes open. Do you find it difficult now to locate the exact position of your partner's finger?

It is clearly much easier to judge distance using both eyes. Have you thought how difficult it would be to play a ball game if you had only one eye? We use both our eyes to locate exactly where the ball is. Wild animals which prey on others for food have both eyes directed forwards so that they can judge exactly the distance to their prey.

Another little experiment very much like the above is to close one eye and then hold out the body of a pen at arm's length. Now with the other hand try to fit the cap on the pen.

Experiment 11.10
Distinct images with two eyes

Keep both eyes open during this experiment. Hold up a pen at arm's length and consciously focus on this. Now place the forefinger of the other hand between the eyes and the pen. What do you see when you are still focusing on the pen? Do you see two images of your forefinger?

Experiment 11.11
Overlapping images

Again keep both eyes open. Now stretch out both of your arms in front of you at face level. Focus on the wall of the room in the background. Now bring the first fingers of both hands towards each other until they almost touch; be careful still to focus on the background. Now separate the fingers. When the fingers are about 2 cm apart, do you find a mysterious floating extra finger appearing?

In Fig. 11.13, focus on the cyclist with the left eye and on the lorry with the right eye. Now bring the page slowly nearer your face until it is almost touching it. What does the cyclist appear to be doing? Does he appear to move to the right?

Experiment 11.12
Range of vision

Again focus on the classroom wall and ask your partner to hold a pen in front of you and then move it gradually to one side. Ask him to stop moving the pen just when you lose sight of it. Now turn your head to see where the pen is. This will give you an idea of how far you can see 'out of the corner of your eye' when you are looking straight forward. Repeat the experiment with the pen moved to the other side, then upwards and downwards. You will find that when looking to the front we are very limited regarding what we can see to the side and above.

Experiment 11.13
Range of colour vision

Again focus with both eyes on the classroom wall and ask your partner to bring forward from behind you different coloured cards, each time sliding the cards forward along the same imaginary line to your front.

Do you begin to see all the coloured cards at the same point? Are you always able to say what the colour is when you first see the card?

We find that we are not usually aware of colour at the edge of our view. Did you find that yellow was the most quickly seen? Most people do. What colour of car do you think a careless motorist, not paying very good attention to traffic passing him, would be aware of first? Can you think which vehicles could most suitably be painted yellow? Is that their colour at present?

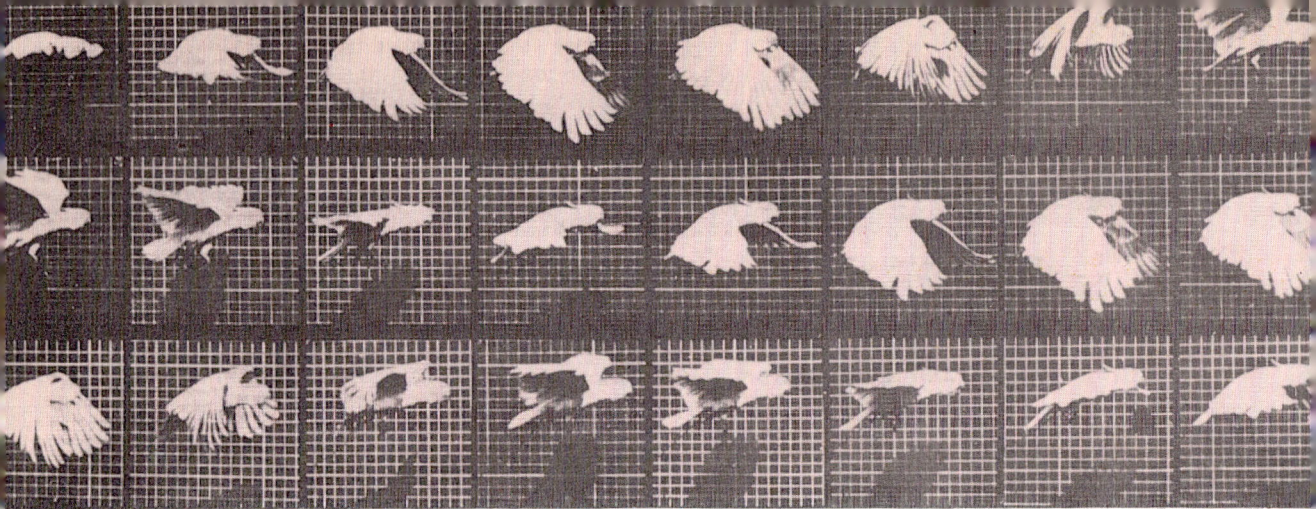

Fig. 11.14 A cine-film sequence showing the movements of a parrot in flight

Experiment 11.14
Persistence of vision

Take a sheet of stiff drawing paper about 15 cm square. On one side draw a bird or an animal, and on the other side draw a fairly large cage with not too many bars (Fig. 11.15). Now take a fairly stiff piece of wire – a long knitting

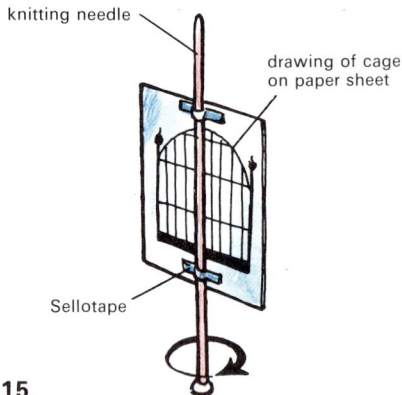

Fig. 11.15

needle will do – and stick this down the centre of one side of the paper. Now spin the needle very quickly between your hands. What do you see? Has the bird or animal got intó the cage?

If you have an old book, make a series of 'match-stick men' drawings in the top corner of each page. Each time put the limbs of the man in a slightly different position. When you flick the pages over you will get the impression that the man is moving. With a little practice you can draw pictures of cats chasing mice, men running and jumping, and so on, which will seem to move smoothly when you flick over the pages.

Have you ever seen a reel of cine film? Something like twenty-four pictures are flashed on the screen every second, but we do not see these as twenty-four still pictures every second. Our eyes blend one image into the other as the nerves on the retina retain the light signal for a short time. If you look fixedly at a lamp and then switch it off you will sometimes see a faint image of it even after the lamp is out. We call this 'persistence of vision'.

Experiment 11.15
The blind spot

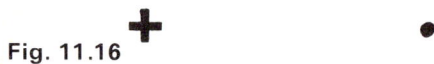

Fig. 11.16

In this experiment cover up your left eye with your left hand, and then look hard at the + above. Now bring your head slowly towards the page. When your eye is about 15 cm away from the page something strange happens. The spot disappears. So we are actually blind at that place. You have found your blind spot for that distance. You will also have a similar blind spot for your left eye. This means that when we look

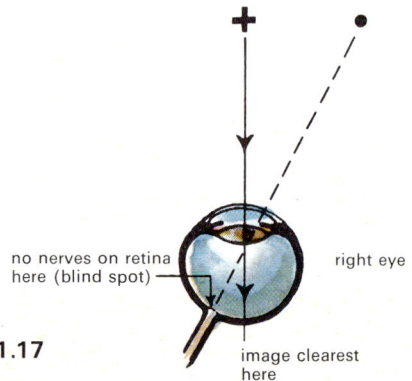

Fig. 11.17

to the front there are two regions, one to the right and the other to the left where we are actually blind, and we would not see anything coming towards us from these directions. This is often important. What do you see a boxer doing so that he is not hit by a punch from his blind direction?

This spot in our eyes is blind because there is no retina at that point. It is where the optic nerve enters the retina.

Experiment 11.16

Look back at the diagrams in Unit 1, page 3. Obviously we cannot always believe exactly what we see. This is because the brain sometimes does strange things with the image formed on the retina. For example, instead of there being holes in what we see, our brain fills up these empty spaces in our vision with parts of neighbouring objects.

Window dressers and shopkeepers must be careful how they display their goods and materials. Put a piece of red paper so that it overlaps a piece of green paper. What do you notice about the join between the two colours? Most people see the colours separated by a thick black line. Is the line in fact there? Try with a number of sheets of different coloured paper, overlapping them as above.

11.11 THE EAR

We are now going to investigate another of our senses – that of hearing. The organ through which we perceive sounds is the ear. Look at the diagram of the ear in Fig. 11.18. You will see that there are three regions.

(a) The outer ear is a flap to catch sound waves and a tube into which the waves are guided on to the ear drum. Some animals can move the ear flaps to pick up sounds from different directions; you have probably noticed your dog or cat doing this. Human beings cannot generally do this, although some of us can 'waggle' our ears. How many ear-wagglers are there in the class?

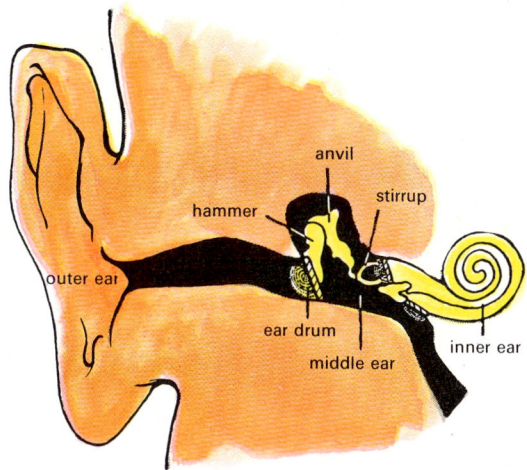

Fig. 11.18

(b) The middle ear. This area acts as a kind of lever. Look at Fig. 11.19. You will see that a

Fig. 11.19

small movement of the boy's finger results in a big movement of the other end of the lever. In the same way, the bones in the middle ear serve to make the vibrations of the ear drum caused by the sound falling on them bigger before they are handed on to the inner ear. In other words they act as an amplifier. These three bones are the smallest in the body.

(c) The inner ear looks like a snail's shell. It contains nerves which each pick out different sounds.

Deafness can result if any of these three parts of the ear are damaged.

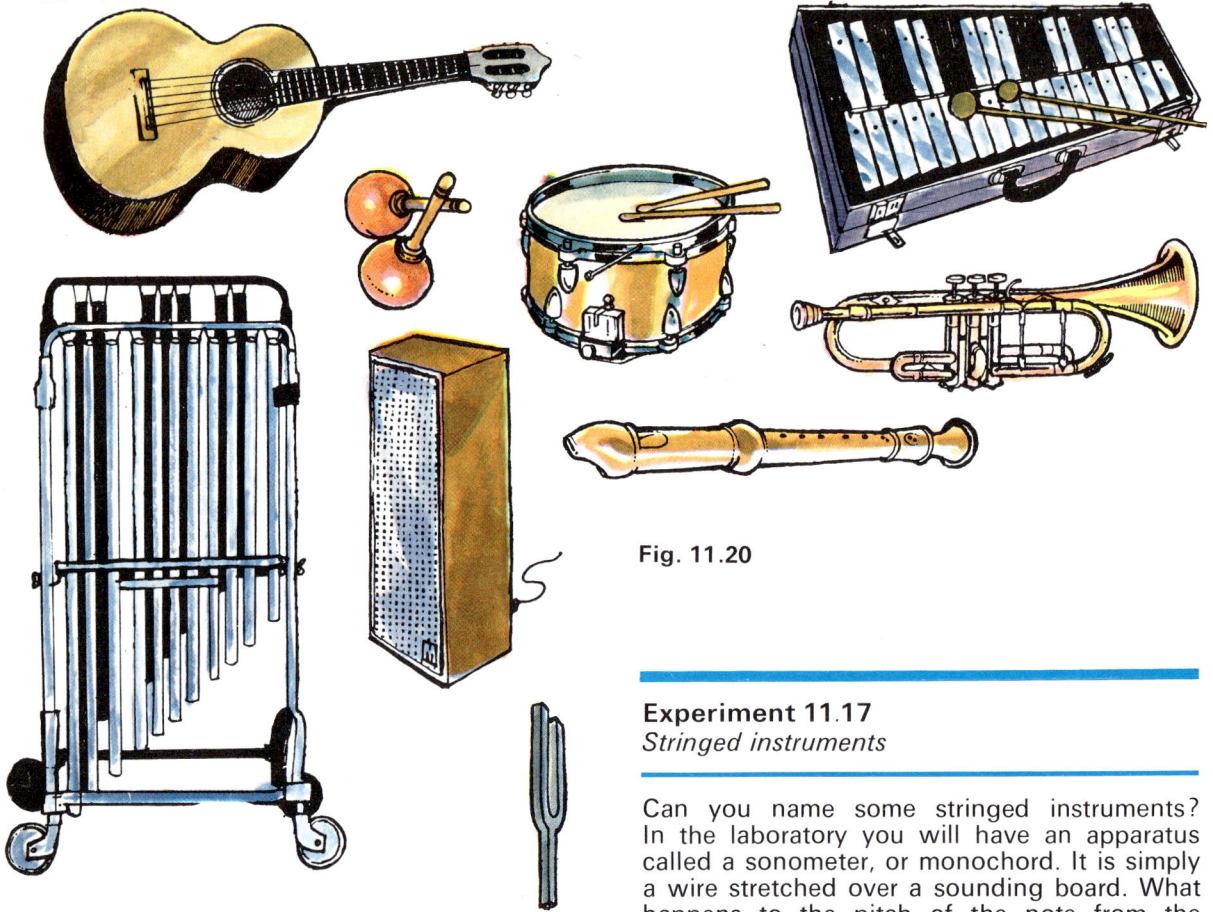

Fig. 11.20

11.12 SOURCES OF SOUND

Each member of the class should bring along to school an example of one means of producing sounds. Exchange them among the members of the class. You might also have some of the apparatus illustrated in this section. Try out as many items as you can, but do not make too much noise all at once! Lightly touch the tuning fork, the loud-speaker, and so on, when they are producing sounds. What are they all doing when they are giving out sounds?

Sounds are produced by things vibrating. When they vibrate they push the air away in front of them, and then pull it back, setting up waves in the air, in just the same way as the vibrator produces waves in a ripple tank. These pulsations in the air reach our ears, where they set the ear drum vibrating. When your teacher is speaking to the class what is being transmitted from him to you? Where did it come from?

Experiment 11.17
Stringed instruments

Can you name some stringed instruments? In the laboratory you will have an apparatus called a sonometer, or monochord. It is simply a wire stretched over a sounding board. What happens to the pitch of the note from the vibrating string when you

(a) tighten the string;
(b) shorten the length of the vibrating part of the string by pressing on it?

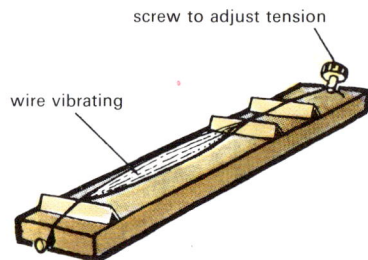

screw to adjust tension

wire vibrating

Fig. 11.21

These two operations will give you some idea of how a stringed instrument is tuned, and can give a range of notes.

If members of the class play stringed instruments perhaps they will bring them to school and you can see how they are tuned, and how they are played.

Fig. 11.22 The strings section of the Orchestra

Experiment 11.18
Your 'voice box'

How do human beings produce the sounds we call speech? Where do we make these sounds? Tilt back your head and stroke a finger tip along the front of your neck, and you will feel your Adam's apple, or larynx. Inside are two flaps of skin (called vocal chords) which we can make vibrate by exhaling air over them. This is very much like the 'squeaker' you can make by stretching a piece of rubber balloon over the end of a tube and blowing down it.

How do you think we raise the pitch of our voice? Speak or sing into a microphone attached to a cathode ray oscilloscope, as you did in Unit 1, and you will see a picture of the sound waves produced by these vibrations.

vocal chords

larynx

Fig. 11.23

Experiment 11.19
The tuning fork

Fig. 11.24

Hold a ruler down firmly with part of it over the edge of the bench. Twang the ruler and notice the pitch of the note. Now change the length of the vibrating part of the ruler. What happens if you shorten this length? Does this agree with what you found for strings?

A tuning fork is rather like a ruler that has been bent into a U shape. Hold a tuning fork by the shaft and strike it against a pad, or against the heel of your shoe if this is rubber. Let the tip of the prong just touch the surface of some water in a basin. What happens to the surface of the water? Does this confirm our idea of what sound is?

Experiment 11.20
Wind instruments

pupil blowing across mouth of bottle

(air vibrating)

BLOGGS DAIRY

Fig. 11.25

Blow across the top of a bottle. Now add a little water so that there is less air in the bottle. Does this air produce a sound, and is its pitch the same as before? You might be able to make a musical scale if you can get a set of eight test-tubes and fill them to different heights with water. The shorter the length of the air, the higher the pitch of the note produced by the vibrating air.

This is the way in which the group of instruments called 'wind instruments' work. Some members of the class may play wind instruments, and could bring them in to school. Perhaps someone will demonstrate how a recorder is played. When you cover up all the air holes it is the whole length of air inside the tube which is vibrating, but when you remove a finger, or fingers, the length of the air which is actually vibrating is shortened, and the pitch of the note rises.

Experiment 11.21
The signal generator

In this experiment you will have an electronic signal generator which is connected up to a cathode ray oscilloscope (c.r.o.) and to a loud-speaker. When the knob of the generator is turned clockwise, the pitch of the note increases. What will happen to the rate of vibration? We often call the number of vibrations in one second the 'frequency' and each vibration per second a "Hertz" (Hz). The signal generator can produce a very wide range of frequencies —much wider than any of the types of instruments you have come across earlier in this section. As the frequency increases what do you notice happening to the length of the waves you see on the c.r.o.? Increasing the frequency should have the effect of making all the waves crowd together so that each complete wave is shorter. The distance between one wave and the next, where it exactly repeats itself, is called the **wave-length**. If you turn up the volume control of the signal generator, what happens to the height of the waves? The height of the waves is called their **amplitude**.

The principle of the signal generator is used in constructing electronic organs. By combining waves together it is possible to make sounds like those produced by any kind of musical instrument.

Experiment 11.22
Your range of hearing

Do you think you can hear the whole range of notes the signal generator can produce? Find out by adjusting the volume control to a suitable level, and putting the frequency controls to the bottom of the scale, then slowly increasing the frequency. What do you notice on the c.r.o. screen?

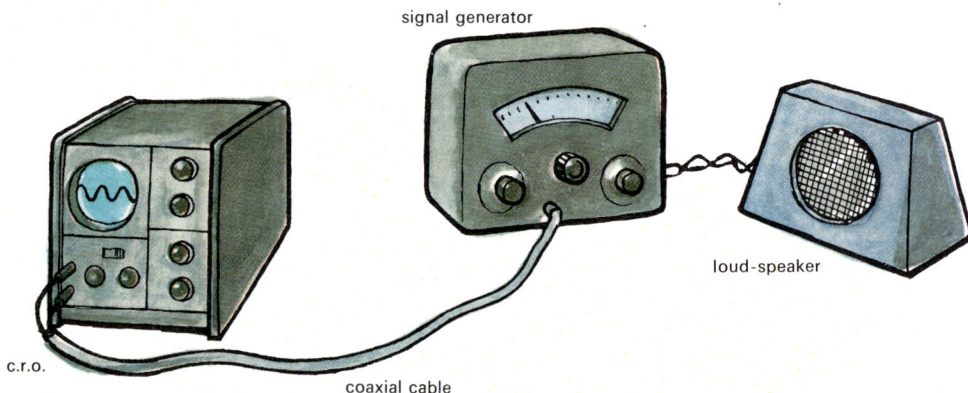

signal generator

loud-speaker

c.r.o.

coaxial cable

Fig. 11.26

Fig. 11.27 The wind/brass section of the Orchestra

Although there are waves to be seen, and the loud-speaker cone can probably be seen moving slowly backwards and forwards, you will probably not be able to hear the very low sound produced, except perhaps as a series of separate plops. As you increase the frequency there will come a time when you can just hear a note. What is the frequency of this, the lowest note you can hear?

Now try to find the frequency of the highest note you can hear. To check up on the fact that there really are vibrations coming from the generator, change the time base control of the c.r.o. to show these waves on the screen. Repeat the experiment, setting the signal generator at its highest frequency on the dial. This will give a note of too high a pitch for you to hear. Now gradually reduce the frequency until you hear a very high-pitched whistle. Compare this frequency with that of the note you just lost when you did the first part of the experiment raising the frequency.

Perhaps you will be able to do this experiment with all the pupils in your year. It is found that all people do not have the same range of notes which they can hear. It is quite likely that some pupils in the group will have a cut-off at say, a frequency of 16 000 Hz, while others can still hear a note at 19 000 Hz. The older one becomes the lower is the cut-off. Older people might not be able to hear the squeak of a mouse, whereas you can. Many animals, such as dogs, are able to hear notes of much higher pitch than we can. To many animals their ability to hear is very important for their own safety. Try to find out how a bat uses its sense of hearing to enable it to fly safely in darkness.

11.13 TRANSMISSION OF SOUND

Experiment 11.23

Fit up the apparatus shown in Fig. 11.28. When the current is switched on, not only

To LV 6 V

Fig. 11.28

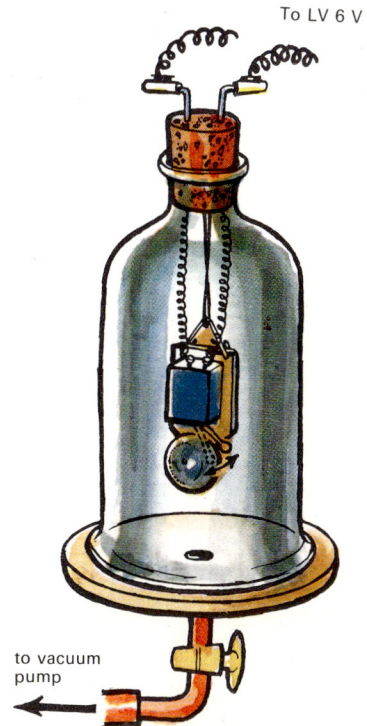

to vacuum pump

should you see the hammer of the bell vibrating, but you should be able to hear the sound given out by the gong. If now the vacuum pump is switched on so that it begins to remove the air from the bell jar, what do you find? Does the hammer still vibrate? What happens to the volume of sound you hear? If you have a good pump it should be able to remove practically all the air. When this happens can you hear any sound at all?

Now switch off the pump and slowly let air back into the bell jar. What happens to the volume of sound?

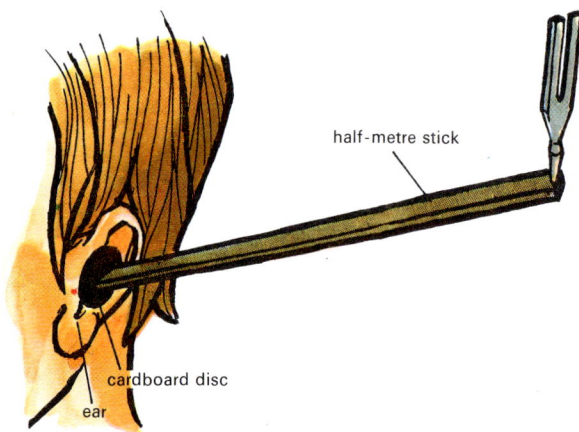

Fig. 11.29

Obviously air is required for the sound vibrations to be able to reach you. This is clearly different from light, and from heat, waves. They can reach us from the sun, and in doing so have to pass through empty space. There may be mighty explosions taking place on the sun, but the noise of them would never reach us because there is a vacuum between us and the sun and sound cannot travel through a vacuum, as our experiment has shown us.

It is interesting to think about the moon which has no atmosphere round about it, so that it exists in total silence. Astronauts on the moon have to talk to each other by radio. How could two astronauts on the moon talk to each other if their radios broke down? To answer this question, hold the end of a half-metre stick to your ear and ask your partner to hold a vibrating tuning fork to the other end. What do you find? You should be able to hear the note given by the fork clearly. How do the vibrations reach you? The vibrations

must be able to travel through the solid wood. The astronauts in the above predicament could only talk to each other by leaning towards each other so that their space helmets touched and the sound vibrations could then pass from one to the other.

11.14 SOME THINGS TO DO AT HOME

We have seen that sound vibrations are unable to pass through a vacuum but we can imagine that there is some way in which sound energy can be passed on from one particle to the next in any material. Try these simple experiments at home.

1. Obtain two empty tins and make sure there are no jagged edges where the end has been removed. With a nail knock a hole in the bottoms of the cans and thread a long piece of string between each can, passing through the holes. Tie knots at the ends of the string so that it cannot be pulled through the cans. Now ask someone to hold one tin to his ear; then, keeping the string very tight, speak into the other tin. Keep the string perpendicular to the end of the tin for the best results.

2. Fill a plastic bag or balloon with water. Hold this to your ear and place a watch on the other side of the bag. Can we conclude from this that sound will travel through water? If you are a swimmer you can easily find the answer to this question in the pool.

3. Go to a fence, a railing, or a pipe in a central heating system. Put one ear to the metal and ask someone some distance away to tap or rub the metal with a screw-driver.

4. Uncoil a hose-pipe and place one end to your ear while someone whispers into the other end. The hose should be stretched to its fullest extent and there should be no kinks in it.

Before the telephone was invented people often used 'speaking tubes'. What you have just been doing is the same sort of thing. In all these experiments, how does the sound energy reach your ear?

11.15 TASTE, SMELL, AND OTHER SENSES

By now you will realize that our senses are limited and we can be easily fooled into seeing things that are not really as we see them. We cannot hear certain sounds, although other creatures are able to. Can we rely on our other senses? Does the entire surface of our tongue detect the taste of all we eat? Do we use our noses only to smell or do we use them to help us to taste? Can we detect pressure, heat, and cold on all parts of our skin? These are the problems you are now going to investigate.

11.16 THE SENSE OF TASTE

Experiment 11.24
Can you detect different tastes?

Liquids A, B, C, and D each have a different taste. They will be either sweet, sour, salt, or bitter. Using a clean glass rod put a drop of liquid A on to your tongue. What taste does it have? Rinse out your mouth with water and in a similar way taste the other liquids in turn.

You should have been able to detect each taste quite easily. But can you detect each on all parts of your tongue or are some regions more sensitive to certain tastes than others? Let us find out. For this experiment work with a partner.

Experiment 11.25
More about taste

Ask your partner to sit with his eyes closed and his tongue extended. Using a clean glass rod, take a drop of one of the liquids used in Experiment 11.24, but do not tell your partner which one you have chosen. Place the liquid on the front of his tongue, and ask him to identify the taste while his tongue is still extended. Record the results in the table below. Test the front, side, and back of the tongue with each liquid.

Region of mouth	Sweet	Sour	Salt	Bitter
Front				
Side				
Back				

Remember to rinse out the mouth before adding a different liquid.

Enter a tick in the appropriate space if he identifies the taste correctly, and a cross if he fails to identify it.

Draw an outline of your tongue and on it outline the regions where each taste could be detected.

Experiment 11.26
Taste and smell

For this experiment use cubes of apple, onion, potato, and turnip. Ask your partner to close his eyes and extend his tongue. Hold a piece of onion to his nose and place a cube of any other food on his tongue. Can he identify it? Place a clip on his nose and then lay cubes of food on his extended tongue. Can he identify them? Does your nose help you to taste?

11.17 THE SENSE OF TOUCH

Experiment 11.27

Move the points of a pair of dividers until they are about 1 cm apart.

Ask your partner to sit with his eyes closed and his hand extended. Lightly touch the surface of his hand with both points of the dividers. Can he feel both points? Touch various parts of his hand and arm, sometimes with one point, sometimes with two, and each time ask him to tell you how many points he feels. Draw an outline of the hand and mark with dots the regions where the points could be felt. Is the entire surface of his skin sensitive to touch?

Experiment 11.28
Temperature and touch

Draw a 2 cm square on the back of your partner's forearm. Remove a mounted needle from either a beaker of hot water or a beaker of ice. Dry the needle and lightly touch the skin within the square. Ask your partner if he feels heat, cold or just touch. Touch different points within the square with hot and cold needles and then map the results as you did in Experiment 11.27.

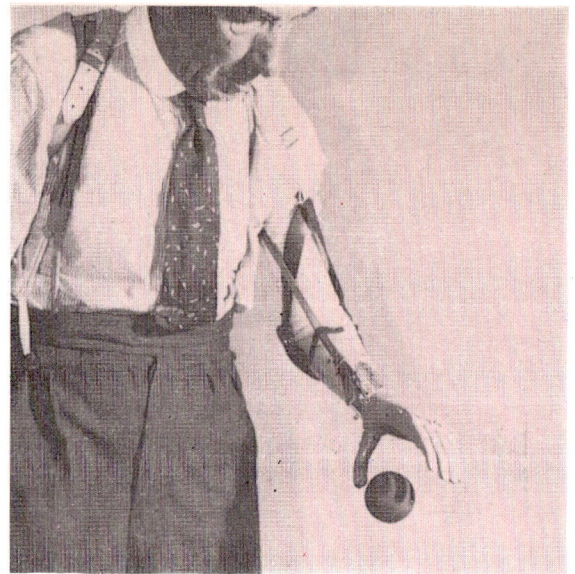

Fig. 11.30 Movements with an artificial arm. Catching a ball

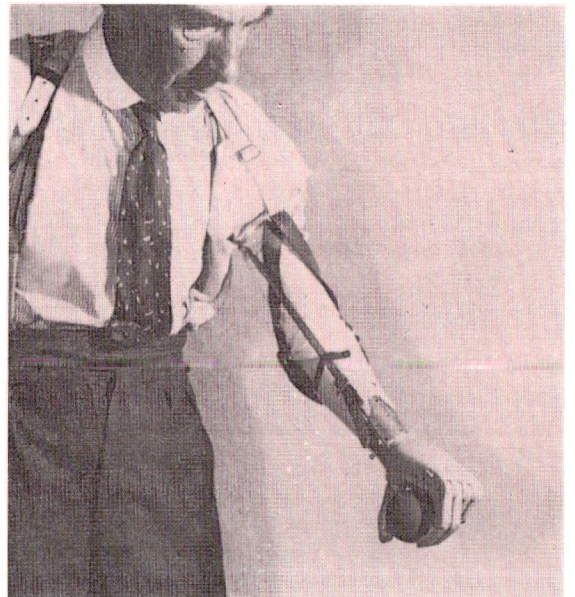

11.18 NERVES

Look at the maps you have drawn showing which parts of the skin were sensitive to touch, to heat, and to cold. Each dot indicates the position of a **nerve** ending. Nerves are collections of special cells which carry information from our surroundings to our brain and spinal cord. Some of this is interpreted by our brain and as a result other information can be carried from the brain by different nerves to muscles and glands of our body which react in a certain way.

Our actions can be divided into two groups, conscious action and reflex action. Our **conscious actions** are under the control of our brain. They occur as a result of thought. We decide to do something e.g. pick up a sweet, and the action is controlled by the brain. Information is sent from the brain to the muscles in our hand and arm which move in such a way that the sweet is picked up.

Certain actions, however, take place automatically, without any thought. These are called **reflex actions**. Reflex actions take place very rapidly and they can help us to escape injury. They make us blink our eyes if bright light is suddenly shone on them, they make us pull our hand away from something hot, they make us sneeze if something irritates the nose.

Experiment 11.29
A reflex action

Sit on a stool with one leg crossed over the knee of the other. Ask your partner to tap the crossed leg just below the knee with the edge of his hand. What happens?

Ask your partner to tap your knee again, but this time try to prevent your foot jerking upwards. Can you do this?

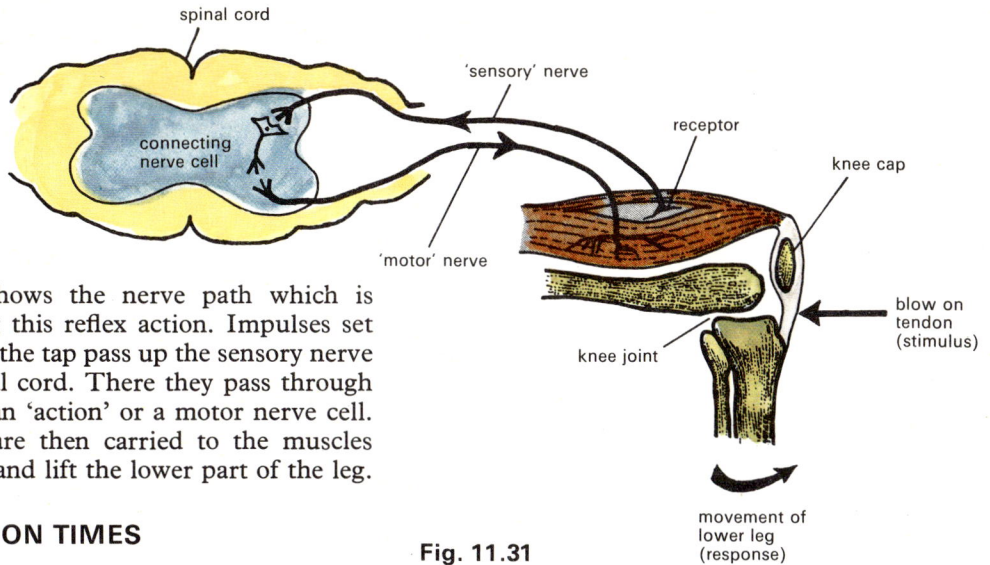

Fig. 11.31

Fig. 11.31 shows the nerve path which is followed during this reflex action. Impulses set up as a result of the tap pass up the sensory nerve cell to the spinal cord. There they pass through a nerve cell to an 'action' or a motor nerve cell. The impulses are then carried to the muscles which contract and lift the lower part of the leg.

11.19 REACTION TIMES

Have you ever heard anyone say 'I have done that so often that I can do it without thinking'? If we repeat a particular action many, many times we can become faster at it, and our brain may not be required in controlling the action. It becomes a reflex action.

Experiment 11.30

All stand in a circle and hold hands. Appoint someone to hold a stop watch. This pupil, A, must start the stop watch with his right hand and at the same time squeeze the right hand of the person standing on his left, B. When B feels his hand squeezed he should then squeeze the right hand of the person on his left, C and so on. Meanwhile A must transfer the stop watch to his left hand. When A's right hand is squeezed the stop watch should be stopped.

How long did it take for the 'squeeze' to pass round the circle? Repeat this several times and note the time taken. Is the reaction time affected by practice?

11.20 THE BRAIN

Fig. 11.32 shows the outline of the human brain. The brain, which contains millions of nerve cells, is divided into three main regions, the fore brain, the hind brain, and the brain stem. The fore brain is the largest part. Its surface is greatly increased by large numbers of folds and wrinkles. This is the part of the brain which controls our conscious

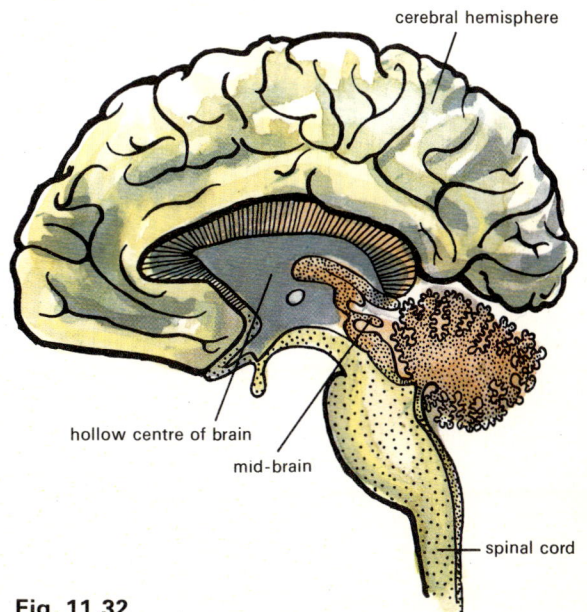

Fig. 11.32

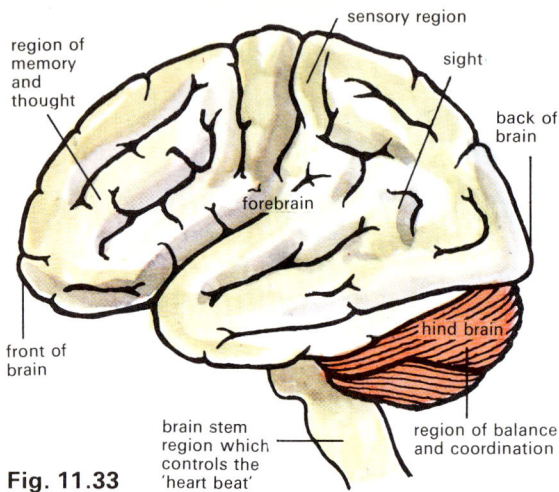

region of memory and thought

sensory region

sight

back of brain

forebrain

front of brain

hind brain

brain stem region which controls the 'heart beat'

region of balance and coordination

Fig. 11.33

actions. Look at Fig. 11.33 and find the part of the brain which

(a) controls memory and thought,
(b) receives information from the eyes and the ears,
(c) coordinates movements.
(d) controls involuntary movements such as the beating of the heart.

WHAT YOU HAVE LEARNT IN THIS UNIT

1. Our eyes are very wonderful instruments and must be looked after very carefully.

2. The convex lens in the eye focuses light on to the retina and can change shape to do this. When focusing near objects, it becomes thicker in the middle and its focal length becomes shorter. When focusing distant objects, it becomes thinner in the middle and its focal length become longer.

3. A concave spectacle lens is required to help people with short sight. A convex spectacle lens is required to help people with long sight.

4. A camera works rather like an eye, but to focus near objects on to a film we screw the lens outwards. To focus distant objects we screw the lens inwards.

5. In both an eye and a camera the image is inverted and reversed.

6. About one male in twelve is colour blind to some extent, and has difficulty in distinguishing between red and green.

7. To judge distance, two eyes (binocular vision) are better than one eye (monocular vision).

8. Our eyes can often deceive us or play tricks, but this is due to the brain not interpreting messages correctly.

9. Sounds are produced as the result of vibrations.

10. The smallest bones in our bodies are in the middle ear and their job is to amplify sound vibrations.

11. Low-pitched sounds are due to slow rates of vibration. High-pitched sounds are due to fast vibration rates. The number of vibrations per second is called the frequency of the vibration and is given as so many Hertz (Hz).

12. The range of pitch of sounds we hear is limited and becomes less as we grow older.

13. Long vibrating wires and air columns produce low-pitched notes. Short vibrating wires and air columns produce high-pitched notes.

14. Sound vibrations require particles to pass them on. These vibrations can pass through solids, liquids, and gases but not through a vacuum.

15. Different regions of the tongue detect different tastes.

16. The senses of taste and smell are closely connected. One affects the other.

17. All areas of the skin are not equally responsive to touch.

18. Messages are taken to the brain by the nerves. Information thus received is interpreted by the brain, and messages are sent along other nerves, causing a response in the muscles and glands.

19. Actions controlled by the brain are called **conscious** actions. There are other automatic responses which are called **reflex** actions.

20. The time that elapses between the receipt of a stimulus to the resulting action is called the reaction time.

Unit Twelve
The Earth and What We Get From It

12.1 THE IMPORTANCE OF THE EARTH

In this unit we are going to study the planet on which we live, and how man has been able to extract from it useful substances which have made life easier for him. Have you ever thought that the earth is our only source of metals, our only source of fuels, our only source of food? It is ultimately the only source of all the chemicals we use in our daily life – our drugs, our plastics, our building materials, everything.

12.2 HOW IT ALL STARTED

As, of course, no-one was alive on the earth until a very long time after it was formed, no-one knows just how it all started. All we can do to find out about this is to use the evidence we have round about us, and make guesses, just as we did when we thought out the nature of matter in Unit 4.

EXPERIMENT 12.1
Looking at rocks

Examine a piece of quartz, a piece of granite, and a piece of calcite. What do they all have in common? Examine the hardness of these rocks. Can you scratch them with a knife? Can you crush them with a hammer? Are some more easily crushed than others? Do they dissolve in water?

We have learnt in Unit 5 that crystals can be produced in at least two ways – one by the cooling of molten substances (you will remember the crystals of metal in the broken door-knob, and the formation of crystals from molten salol), and another by the cooling of saturated solutions (for example alum and potassium nitrate). It seems very likely that the quartz, granite, and calcite were formed in one or both of these ways.

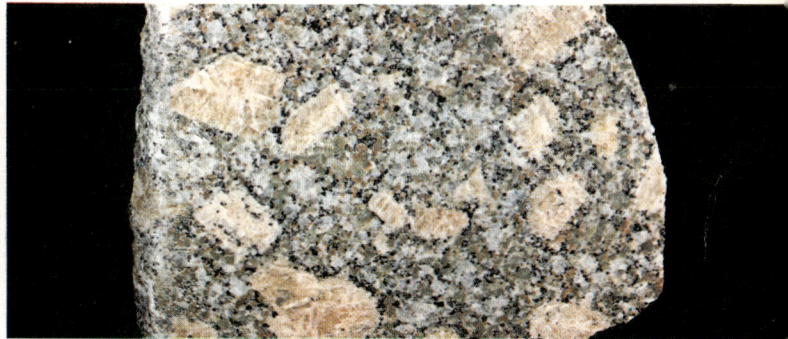

Fig. 12.1 Crystals of (*above*) quartz, (*right, top*) granite and (*right*) calcite

As you have found that these rocks do not apparently dissolve in water, what is the most likely way in which they were formed?

By looking at these rocks we have been able to come to the conclusion that the earth was at some time in the long distant past in a molten state. The outside cooled and the molten rocks became solid. This idea is supported by the fact that the interior of the earth is, even now, much hotter than the outside layer. The temperature in a coal mine is higher than that at the surface. It is thought that the interior of the earth is still so hot that the material there is molten. This is shown in volcanic districts by the fact that when the volcano erupts liquid rock is often poured out in the form of lava together with showers of ashes and jets of steam and hot water.

Here is a list of some volcanoes. Find out from your atlas where they are:

Etna, Vesuvius, Fujiyama, Krakatoa, Stromboli, Popocatepetl.

There are no longer any active volcanoes in Scotland, but there is much evidence that some of the hills and mountains are of volcanic origin. For example, the rock on which Edinburgh Castle stands is a volcanic plug, and is harder than the rock round about, which has been removed by ice erosion.

Rocks which have been molten at some time but have now set solid are called **igneous rocks**.

At one time in the history of the earth, water was formed and covered a large area of the surface. At the present time about three-quarters of the surface of the earth is covered with water or ice.

Fig. 12.3 Edinburgh Castle which is built on a volcanic plug

Although rocks may not be soluble in water, nevertheless the wind, rain, and frost may break them up into smaller pieces. These are washed down by rivers and eventually reach the sea – maybe after many thousands of years – as mud, silt, and sand, which accumulate at the bottom of the sea, or in lakes. As the deposit gets thicker, the bottom part is squeezed more and more, and becomes a compact mass. Often the particles are actually cemented together through substances produced by chemical reactions. The shells of dead sea-organisms, which are made of calcium carbonate (or chalk), may form a layer on top of the mass, or at intervals between layers. Then the sea may have receded, or earth movements may have taken place, making the sea bed dry land. What *was* the sea-floor may now be hills or even mountain ranges. Rocks of this kind are called **sedimentary** rocks, and include limestone, chalk, sandstone, and shales.

Fig. 12.2 A volcano during eruption

Fig. 12.4 Sedimentary rocks in Alum Bay on the Isle of Wight

Experiment 12.2
Sediments

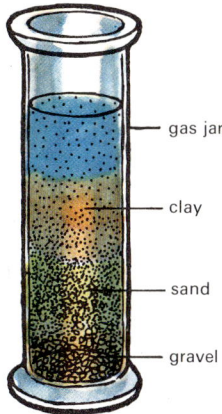

Fig. 12.5

(a) Three-quarters fill a large gas jar with water. Mix some fine gravel, sand, and powdered clay together, and add this to the water in the gas jar until the latter is nearly full. Leave to settle, and note all that happens.
(b) Repeat using salt water.
(c) Repeat again, swirling the water after adding the sediments.
What is likely to happen when muddy water reaches the sea?

Sedimentary rocks may be recognized by the fact that they are frequently made up of rounded particles, are often layered, and contain fossils (the remains of once living plants and animals) or the imprints of them.

Collect as many different samples of sand as you can. Notice their colours and examine them under a hand lens. Try to account for the differences between them.

Geologists talk of another kind of rock called **metamorphic rock**. The word 'metamorphic' means change, and these rocks have been produced from already existing ones by the action of heat and/or pressure. They may have been crushed, baked, chemically changed, and perhaps melted. The picture (Fig. 12.6) shows a typical metamorphic rock. Here are the names of some: slate, schist, gneiss, quartzite, marble.

The last one is a form of calcium carbonate which was probably once chalk or limestone, made up of the shells of dead marine organisms. It has been subjected to heat and pressure and converted into a crystalline form.

Fig. 12.6 Marble – a metamorphic rock – being cut from mountains in the Apuan Alps in Italy

It is sometimes difficult to tell whether a rock is igneous or metamorphic. Why? Which kind of rock – igneous, sedimentary, or metamorphic – do you think will be the oldest and which the youngest?

12.3 THE STRUCTURE OF THE EARTH

Fig. 12.7 shows what scientists believe the structure of the earth to be like. The earth is more than 6000 km in radius. The outer crust is very thin, only about 48 km deep; below it is a 'semi-plastic' mantle, followed by a liquid outer core, then a solid inner core. The pressure at the surface of the outer core is about 1.75×10^6 atm and at the surface of the solid core about 4×10^6 atm. We write atm to represent 1 atmosphere as the size of the air pressure at the earth's surface. These are enormous pressures which we find it very difficult to imagine. The temperature of the solid core, which is thought to be made up mainly of iron, is 3000–6000 °C.

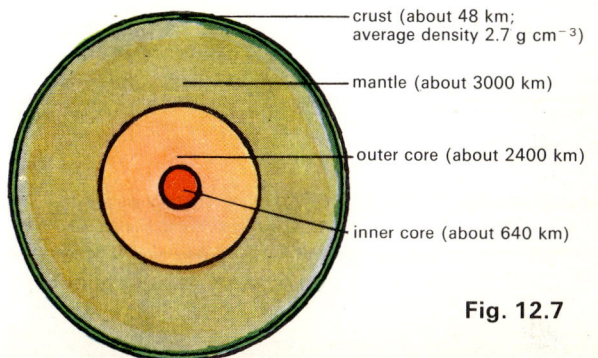

crust (about 48 km; average density 2.7 g cm^{-3})

mantle (about 3000 km)

outer core (about 2400 km)

inner core (about 640 km)

Fig. 12.7

Some questions for you to answer

1. Why is the inner core solid, when the normal melting point of iron is 1500 °C?

2. Why is the density of the inner core about 12 g cm^{-3} whereas the density of iron is 7.9 cm^{-3}?

Experiment 12.3

Find the densities of a number of different rocks. Different groups might take different specimens. How does the average result for the class compare with the density of the crust? Is there any connection between the type (igneous, metamorphic, sedimentary) of rock and its density?

12.4 ELEMENTS IN NATURE

What elements occur naturally? Make a list of those you know of, but do not include any you are not sure about. Does your list contain the gases oxygen, nitrogen, and helium and the other noble gases, or did you forget these? What about iron? Why did you not include it?

Actually there is a little iron in a few places, but it has all come from meteorites, perhaps many thousands of years ago. Why hasn't it rusted? The reason is that it is not often pure iron, but contains traces of other elements such as nickel, manganese, and chromium, which have, in effect,

made it into stainless steel. There is also a very little copper, but this is so rare as to be negligible as a source of the metal.

Did you remember to include sulphur and gold? Some others that you might not know about are mercury, silver, and platinum, but again the quantity of these is small.

Would you expect to find any hydrogen in the atmosphere of the earth? Why?

Something that needs to be explained is why so few elements out of the ninety-two occur by themselves. The following experiments may help us.

12.5 METALS AND SULPHUR

Experiment 12.4
The action of sulphur on metals

(a) Mix a little of the following metals with some powdered sulphur (flowers of sulphur will do) and heat the mixture gently in a small test-tube. Notice whether any reaction takes place. The metals to use are iron (filings), zinc foil, copper powder, lead shot, aluminium foil, and magnesium ribbon (in small pieces). Different members of the class can try different metals.
(b) Get a small piece of silver foil and rub a little powdered sulphur on it.
(c) Rub a small drop of mercury with a little powdered sulphur in a mortar, using a pestle.

Notice if any new substance is formed and whether the reaction is vigorous or not. Try to arrange the metals in order of activity. This cannot be done very accurately because the rate and vigour of the reaction will depend on how well the metal and the sulphur can be mixed, i.e. on how finely powdered they are. Why should this make a difference? Do you remember doing two of these experiments before? Look up Unit 4.

Sulphur is often to be found in volcanic areas, and you will probably have some crystals of sulphur in your school collection of minerals. There is clearly a good deal of sulphur present in the earth's crust.

Can you now understand why there are so many sulphides occurring naturally? Make a list of the sulphides in your collection of minerals.

Fig. 12.8 A sulphuric acid plant in Chile. Piles of sulphur-bearing ore can be seen in the background

12.6 METALS AND OXYGEN

Another very active element surrounds the earth. What is it? You have already done some experiments in Unit 8 on the combination of metals with oxygen. We shall repeat some of these to refresh your memory, and we shall also try some new ones.

Experiment 12.5
The action of oxygen on metals

An easy way of finding out how well metals burn in oxygen is by using the apparatus in

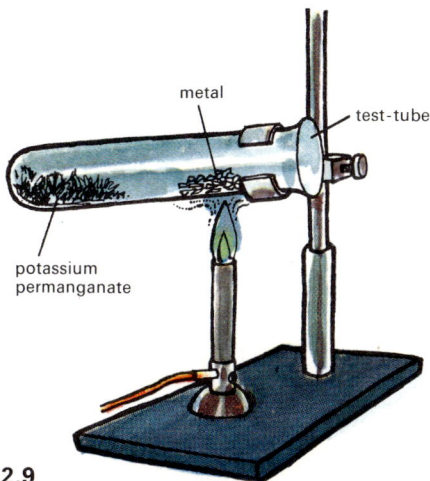

metal

test-tube

potassium
permanganate

Fig. 12.9

Fig. 12.9. When the potassium permanganate is heated it provides oxygen gas. Try heating some by *itself* in a small test-tube and test anything that comes off for oxygen.

In the apparatus in the diagram the oxygen passes along the tube over the heated metal. Note what happens and how vigorous the reaction is. Use the following: magnesium ribbon (NOT powder), calcium turnings, aluminium foil, zinc foil, lead foil, iron filings, copper filings, silver foil.

Can you now explain why *none* of the following metals occurs uncombined in the earth's crust: sodium, potassium, calcium, magnesium, aluminium, zinc, iron (except in meteorites), lead, and why there is very little copper? Why, do you think, is there some silver, mercury, gold, and platinum?

Minerals from which metals are obtained are called **ores**. It is not surprising that many ores are oxides or sulphides. They are, of course, often mixed up with other minerals, and sometimes, before the metal can be extracted from them, they have to be purified. Here are the names of some of them:

bauxite	aluminium oxide
haematite	iron(III) oxide
galena	lead sulphide
iron pyrites	a sulphide of iron
copper pyrites	a sulphide of copper and iron
zinc blende	zinc sulphide

See if these are in your collection.

12.7 CARBONATES

Some minerals, as was explained on page 42, have, in past ages, been washed down into the sea. Sea water and rain water contain dissolved carbon dioxide, and after a very long time these minerals have been changed into carbonates by the action of the water. They have formed layers on the sea bed, and the shells of dead creatures which have lived in the sea have been deposited on top of them. These shells are made of calcium carbonate. Then, perhaps, there has been some movement of the sea bed which has caused the bottom of the sea to become dry land. So we often find carbonate minerals. Here are the names of some of them:

chalk	
calcite	
limestone	all forms of calcium carbonate
coral	
marble	
magnesite	magnesium carbonate
calamine	zinc carbonate
cerussite	lead carbonate
ironstone	a carbonate of iron

Experiment 12.6
Heating carbonates

(a) Heat in separate small test-tubes a little powdered chalk (natural chalk, not blackboard chalk which is not really chalk at all), zinc carbonate, and lead carbonate, and notice everything that happens. Test to see if any gas comes off.

(b) Hold a small piece of marble in a pair of tongs and heat it strongly, then allow the product to cool.

(c) Meanwhile look at some zinc oxide and some lead oxide. Heat a little of the zinc oxide and notice what happens.

(d) Take the product from (b), put it on a watch glass, and add a drop of water to it. What happens?

(e) Take a small piece of calcium oxide on a watch glass and add a drop of water to it.

Some questions for you to answer

1. What happens when carbonates are heated?

2. Can an igneous rock be a carbonate? If not, why?

3. Do all the carbonates you have heated break up equally readily, or are some more difficult to break up than others?

4. Is there any connection between this and the activity series?

5. Can we assume that all carbonates break up when they are heated?

6. Bearing in mind your answer to question 4, predict one carbonate which might be very difficult or impossible to decompose in the laboratory. Try out your prediction by experiment.

Experiment 12.7
More experiments with carbonates

Different groups in the class should each take two of the following carbonates; sodium, calcium, magnesium, zinc, barium, potassium. Add a little dilute hydrochloric acid to each. What happens? Repeat the experiment with dilute acetic acid (vinegar).

All carbonates react with dilute hydrochloric acid and usually with other acids too.

12.8 HOW ARE METALS OBTAINED FROM THEIR ORES?

We have seen that when most carbonates are heated they form oxides and give off carbon dioxide. Many sulphides also form oxides when they are strongly heated in air.

Experiment 12.8
Heating iron pyrites in air

Iron pyrites, as we have said, is a sulphide of iron. Take a little heap of powdered iron pyrites on a piece of asbestos paper, and heat it. Can you smell a gas which is given off? What does the smell remind you of? What does the residue look like?

In many cases, therefore, the production of a metal from an ore is a matter of pulling away the oxygen from an oxide which has been obtained by heating the ore in air. We have already seen in Unit 10 how this can be done. We must look for something which has a stronger pull on oxygen than the metal has. Can you think of one element that might do this?

Another element that can do this is carbon. Carbon, as you know, burns vigorously in oxygen and a great deal of energy is given out. It might be expected, therefore, that heating some oxides with carbon will leave the metal behind.

Experiment 12.9
Heating oxides with carbon

This experiment is best done by heating a mixture of the oxide and carbon on a piece of asbestos paper. Try the substance left from Experiment 12.8, lead oxide, and copper(II) oxide. How will you test to see if any iron is formed? If any lead is obtained from the lead oxide you will see it as shiny globules. How will you know if any copper is formed from the copper(II) oxide? Were you successful?

The removal of oxygen from a substance is called **reduction**, and the oxide is said to have been **reduced**.

Now try a mixture of calcium oxide and carbon, and magnesium oxide and carbon. Were you successful this time?

Some questions for you to answer.

1. Is carbon able to reduce all metallic oxides?

2. Is it likely to be able to reduce aluminium oxide? – or copper oxide?

3. When carbon reduces an oxide, what happens to the carbon?

Fig. 12.10 (*left*) A general view of the blast furnaces and (*right*) molten pig iron being charged at Ravenscraig

The heating of iron oxide with carbon forms the basis of the process used for smelting iron. The iron ore, haematite, which is impure iron(III) oxide, is mixed with coke (a form of carbon) and limestone and heated strongly in a strong current of air in a blast furnace. The limestone removes impurities in the ore. Molten iron is run off into sand moulds and is later converted into steel. In which parts of Scotland is this process carried out?

Experiment 12.10
An investigation of malachite

Malachite is a beautiful green mineral. It is sometimes used in jewellery, but is not really a precious stone. Your teacher will show you a piece. For this experiment you will use the powdered mineral.

Now set to work and see if you can find out what it is. Remember the various ways in which you have examined minerals in the last few lessons. Try to work out a plan of campaign before you start — and work systematically.

12.9 SILICA AND SILICATES

Many minerals contain the element silicon. Silicon dioxide is called silica and sand is an impure form of it.

Fig. 12.11 Mountains and rocks at Loch Corruisk, Isle of Skye. Are these made of silicates?

You have probably seen sand on the sea shore. When you look at it you think that it must be a very stable substance and one that is not attacked by many other things. You are quite right. Silica is one of the most stable substances known, and that is why so many rocks are silicates or mixtures of silicates, e.g. clay, gneiss, slate, and schists.

Experiment 12.11
Some experiments with silicates

Test the solubility in water, the effect of dilute hydrochloric acid, and the effect of heat on some silicate rocks. You might use sand, clay, mica, and felspar. Draw up a table showing your results.

Why are some of the highest mountains in the world made up largely of silicates? Why is it that silicates have been used for a very long time as building materials?

Clay, which is formed by the breakdown of granite by the action of the weather, is a complex silicate which has been used by mankind for ages. Men have built their dwellings out of it – the so-called 'mud huts' of primitive peoples were really made of clay which hardens in the sun to a kind of brick-like material, and of course we still use bricks today to build our houses. They are simply clay which has been heated in a furnace. So, although it is easier to build with bricks because they have a regular shape, we are not much further on than the primitive people who fashioned their houses from wet clay and allowed it to become baked hard in the sun. Clay has also been used by potters to make vessels for holding solids and liquids and for cooking. Cement and concrete are mixtures of silicates.

Fig. 12.12 Making bricks. (a) Digging out the clay, (b) grinding pans reducing the clay, (c) unfired (green) bricks ready for the kiln, (d) bricks packed into the kiln, (e) a batch of finished bricks ready for despatch, and (f) a general view of the kilns and chimneys in the Potteries

Experiment 12.12
Some experiments with clay

Prepare and roll out some potter's clay and cut it into small squares. Have some of them fired in a kiln. Leave others to stand on an open shelf for some time to dry out. Examine a few immediately after they have been rolled. What effect do you think that heat has had on the wet clay?

Experiment 12.13
Glazes

Take one of the fired tiles you made in the last experiment; dip it in salt solution and refire it. Examine the tile on cooling. What has happened? Make a glaze mixture by grinding together 2 g of red lead, 2 g anhydrous sodium carbonate, 1 g of powdered flint (or silica) and 0.5 g of cobalt(II) nitrate. Add water to the mixture to make a smooth paste and paint it on one of the fired pottery tiles you made in the last experiment. Evaporate the water by gently heating, and then fire the tile in the kiln, or heat it in a blowpipe flame. What colour of glaze do you get?

Other members of the class might try a glaze mixture which contains manganese(II) sulphate, iron(II) chloride, chromium(II) sulphate, or nickel(II) sulphate in place of the cobalt(II) nitrate.

A question for you to answer

Clay contains aluminium silicate. Aluminium is a very important metal, yet it is never made from clay which is more abundant than any other aluminium containing mineral. Why not?

12.10 COAL

Coal is the fossilized remains of vegetation which has been changed by the pressure of masses of overlying rock. It probably started like peat, which is found in many British moorlands, and particularly in boggy Scottish hills, where heather has changed into a dark-brown or almost black mass.

Experiment 12.14
Peat

Examine some peat and see whether you can discover evidence that it is formed from decayed vegetation. Try to burn it.

In the Highlands of Scotland, and in Ireland, peat is often burnt as a fuel. This peaty material became covered with rock and soil, and the pressure changed it into more compact material. The nature of the coal produced in this way depends on the pressure to which it has been subjected. Next to peat comes brown coal, sometimes called lignite or cannel coal. The word 'cannel' is a corruption of the word 'candle' and the name was given to this kind of coal because you could light it with a match and it would burn like a candle. If you have a piece of it, try it and see.

The word 'lignite' means 'like wood'. This name too refers to the ease with which this kind of coal can be burnt. There is very little brown coal, or lignite, in this country.

Next in the scale comes bituminous coal. This is the kind of coal which is used mainly for household fires. It does not light so easily as lignite, but is not difficult to start burning as it contains a good deal of tarry matter, and its bright flame is pleasing in an open fire. However, bituminous coal leaves a fair amount of ash.

Next comes anthracite. This is a very hard coal which is difficult to start burning, but once it gets going it gives a very hot flame and little ash. It is almost smokeless. It is used in stoves for heating water in domestic hot water systems, where the heating value of the coal is of greater importance than the appearance of the fire. It is also used in industry to heat boilers to raise steam, for example in running a power station.

To sum up, the various kinds of coal are

peat → brown coal → bituminous → anthracite
       lignite   coal
       cannel coal

As we go from peat to anthracite the coal becomes more difficult to burn, the calorific value (i.e. the heating power) increases, the amount of ash left behind decreases, and the percentage of carbon increases. Look at samples of each of these kinds of coal in your school collection of minerals. Which dirties your hands most?

We all know what happens when coal is heated in the open air because we have all seen coal fires. What happens if the coal is heated in an enclosed space? First, ask yourself if it will burn. Then try the following experiment and see if you were right.

Experiment 12.15
Heating coal

Bring to school a small empty coffee tin and a few pieces of coal small enough to go into the tin and sufficient to fill it a quarter to a half full. Make a hole in the lid of the tin with a nail. Heat the tin and the coal. What happens? Let the tin cool. Take off the lid. What is left inside?

Experiment 12.16
Making coal gas

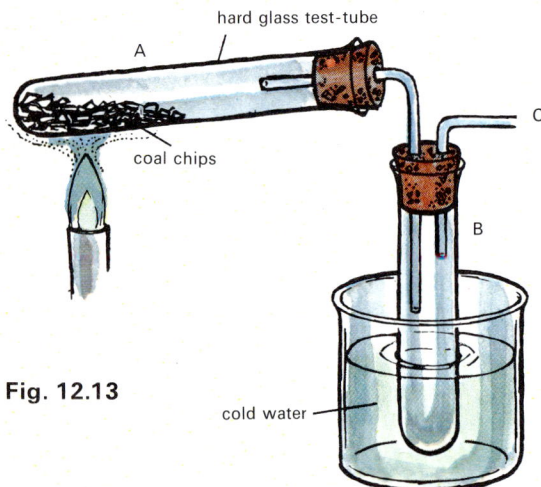

hard glass test-tube
A
C
coal chips
B

Fig. 12.13

cold water

We can see more clearly what happens when coal is heated by using the apparatus in Fig. 12.13. Heat some small pieces of coal in the hard glass test-tube A. What collects in the tube B? Smell it. Test it with indicator paper. Can you burn a gas at the jet C?

The gas obtained by heating coal in this way is called coal gas. Until recently the gas used in cooking stoves, and in some places for lighting, was coal gas, but now it is often mixed with natural gas, or with fuel gases which are the by-products of some industrial processes such as the

petroleum industry. In some places natural gas only is used. So now we call the gas supplied to our houses (or to the laboratory) not coal gas, but 'town gas'.

You will have found a tarry liquid collected in the tube B. This is coal-tar, and from it a large number of useful products can be obtained by distillation. Some of these are benzene, toluene, and naphthalene, and from them many more very important chemicals can be manufactured.

The solid product left behind in the tube after the coal has been heated is coke. In these days coal is very often heated like this just to get the coke and the coal-tar, the gas being merely a by-product. Coke is a very useful smokeless fuel and is used a great deal in the smelting of metals as a reducing agent. It is nearly pure carbon. Look back to page 47, where the smelting of iron was referred to.

Experiment 12.17
Heating wood

If you have time, carry out a similar experiment to Experiment 12.16 but use small pieces of wood instead of coal. Test the products in the same way as before. What is left in the heating tube?

Coal is indeed a storehouse of energy and the starting point for a great deal of our chemical industry. Where did the energy which we can now obtain by the burning of coal, coke, and coal gas come from?

12.11 OIL

The other important fuel obtained from the earth is oil. Unlike coal, it has been formed from the decayed remains of sea animals and plants which settled on the floor of ancient seas. As the mud hardened into rocks, the oily material moved upwards. Some reached the surface, where the portions which boiled at lower temperatures evaporated away leaving a tarry deposit, as in the Pitch Lake in Trinidad. Most of the oil, however,

Fig. 12.14 Boys on the Trinidad Pitch Lake showing the consistency of natural asphalt. Find out how this asphalt is removed and exported

was trapped by hard rock that it was unable to penetrate.

Together with the oil there is always natural gas, which will be on top of the oil itself, and water, which will be below the oil. Why? To obtain the oil it is necessary to bore through the layers of hard rock above the deposits. If the pressure of the natural gas is sufficient, the oil is then blown up to the surface and comes gushing out. In places where there is not sufficient pressure of gas the oil has to be pumped up.

The oil as it reaches the surface does not look very much like the stuff we buy at the garage and use in our cars. To obtain this, the oil has to be refined. Because of the different boiling points of the constituents of the oil, it can be separated by fractional distillation. If you have forgotten about this, refer back to Unit 5 (Book 1, page 77).

The fraction which boils at the lowest temperature contains a high proportion of octane. This is used for making 'aviation spirit' for aircraft, and high-octane petrols for cars. The ordinary grade of petrol contains less octane and more of the fractions which boil at a slightly higher temperature. Why is it easier to start a car in the winter if you are using high-octane petrol than if you use the cheaper grades?

After the petrol range comes the oil used for burning – kerosene (or paraffin); then the light lubricating oil, such as you would use to oil your sewing machine or your bicycle, then the heavier lubricating oil which you use in your car sump, and finally vaseline and wax.

Of course, the composition of crude oil is different in different parts of the world. Some of it contains a lot of heavy oil, and very little petrol. In order to make more of the lighter fractions from this kind of oil, the heavy oil is submitted to a process called 'cracking' whereby the higher-boiling oils are broken down to lower-boiling ones, for which there is a greater demand. You will learn more about cracking later on in your course.

The natural gas which accompanies oil consists largely of methane, a compound of carbon and hydrogen. The discovery of oil deposits below the North Sea, with the natural gas there too, has made it possible to supply a large area of Britain with this source of fuel, and to replace coal gas completely in these parts.

Experiment 12.18
Distilling crude oil

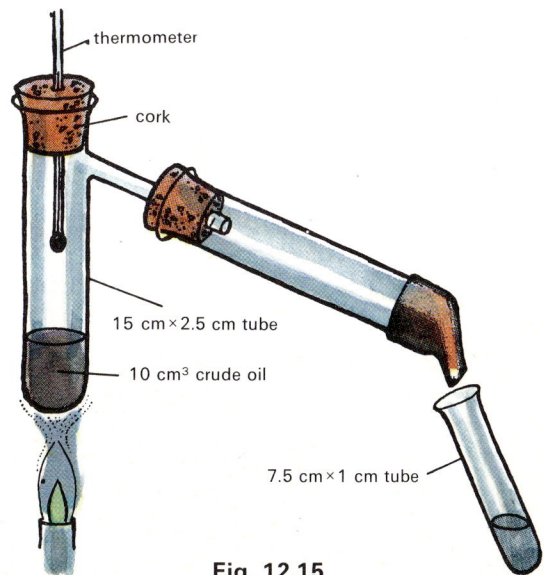

Fig. 12.15

You can imitate in the laboratory what is done in an oil refinery. Set up the apparatus shown in Fig. 12.15. Collect the products of distillation at approximately 0.5 cm^3 intervals.

Smell the products. Find out which of them will burn by pouring some into an evaporating dish and applying a lighted taper. Put down your results in the form of a table.

Fraction	Reading on thermometer (°C)	Smell	Burns (yes/no)
1			
2			
3			
4			
5			
.			
.			
.			
.			

A question for you to answer

1. What would you write in the spaces A, B, and C?

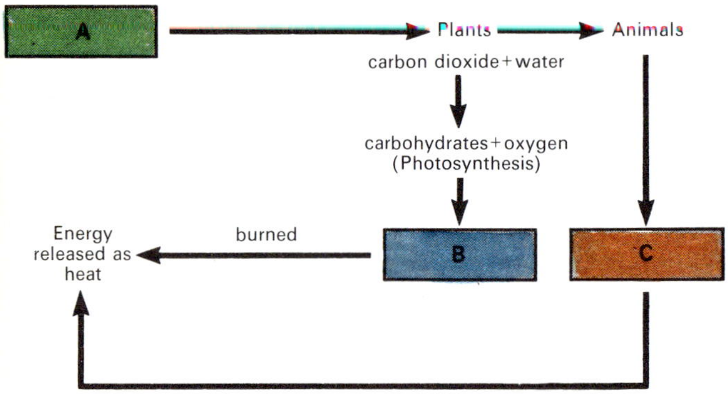

12.12 THE SEA

You have already learnt something about sea water in Unit 5 (Book 1, page 77). We are now going to study this further.

Experiment 12.19
Evaporating water

Evaporate a few drops of distilled water, tap water, and sea water on small pieces of black glass by means of an infra-red lamp. (If you do not have black glass you can use a watch glass and put it on a sheet of black or brown paper.) Which leaves the most residue?

Take about half the residue from the sea water on a clean iron wire, or a pencil lead and hold it in a flame. What happens to the flame? Try the same thing with a drop of sodium chloride solution.

Dissolve the remainder of the residue from the sea water in a little water and add a drop of silver nitrate solution. What happens? Try the same thing with a drop of sodium chloride solution. What do you conclude about the residue from the evaporation of sea water?

The tests you have just carried out are useful ones for sodium and for a chloride. All sodium compounds give a yellow colour to the flame. (You may have noticed this if you have seen salted water, in which potatoes are cooking, boil over.) All solutions containing a chloride give a white precipitate when silver nitrate solution is added. This precipitate dissolves again when ammonia solution is added to it. What do you think the white precipitate is?

Sea water contains many other dissolved substances as well as sodium chloride. As the shells of marine organisms contain calcium carbonate there must be some calcium compounds in the water from which these shells can be made. Why could it not be calcium carbonate itself?

Actually there are considerable amounts of calcium bicarbonate and calcium sulphate in sea water, as well as magnesium sulphate, magnesium chloride, and potassium bromide. At present sea water is the only source of the metal magnesium. Another name for calcium sulphate is gypsum. Perhaps you have a specimen of this mineral in your collection.

All salt deposits have been produced by the evaporation of sea water; so wherever salt deposits are found we may be sure that the area was at one time under the sea. In Britain the largest deposits are in Cheshire, and salt is obtained from them by pumping down water. This forms a large underground lake of salt solution

Fig. 12.16 The effects of erosion on mountains and coasts. How is the erosion caused?

from which the brine is pumped up and then evaporated. The evaporation is done in vacuum pans. This kind of mining is called 'solution mining', and can be used, of course, in comparatively few cases.

Some questions for you to answer

1. When a solution is evaporated in a vacuum will it boil at a lower or a higher temperature than it would under ordinary conditions? Think this out on the basis of the kinetic theory of matter.

2. Will the crystals obtained by vacuum evaporation be large or small?

In Europe, salt deposits are found in Germany at a place called Stassfurt, and in Poland. Here the salt is mined in the usual way and not dissolved in water as it is in Britain.

Salt is the basis of a large amount of our chemical industry. That is why Cheshire and south Lancashire is one of the great chemical industrial areas of Britain.

12.13 THE SOIL ENVIRONMENT

Most of the surface of the earth is covered by a layer of soil. The depth of soil varies in different regions from a thin layer about 0.5 cm deep on mountains to a layer several feet thick in cultivated areas.

The soil is the place or **habitat** where the majority of land plants and many kinds of animals are found. Why is soil the habitat for so many organisms? An organism will live successfully in an environment if the environment provides it with all or most of its needs. Make a list of the requirements that plants and animals need to live successfully. We can find out how the soil provides these requirements by studying its composition.

Experiment 12.20

Place a very small sample of soil on a cavity slide, mix it with water and examine it under the low power of the microscope.

Are all the particles of different sizes or the same size? Are the particles of different or the same shape? Are they all dull or do some glitter or sparkle or appear transparent? Are living things present or absent? Compare soils from different places.

Soil is formed when rocks are broken down into small particles. This is brought about mainly by the action of water, frost, and wind. When water freezes it expands. If rain water falls into the cracks in rocks and freezes, the crack is enlarged and a piece of rock may become separated.

Glaciers remove part of the rocks through which they pass and grind up rocks below them into very fine particles. The sea pounds on rocks and breaks them up. Rivers wear away the rocks through which they pass. Small sand particles carried by the wind bombard soft rocks and break off parts of them.

The broken particles of rock either remain on top of the parent rock or are carried by rivers and glaciers to other regions. Small plants grow on these rock particles, die and form an organic material called **humus**. Soil therefore contains a complex mixture of mineral fragments, of different sizes, and organic material.

We can find out what is present in a sample of soil by separating it into its components. How can you do this? You could spread the sample out on to a sheet of paper and pick out all the plant remains, then the large mineral fragments, then the smaller ones and so on, but that would take too long. There is a quicker method. This method was used by gold prospectors to separate gold particles quickly from the other mineral particles. How did they do this? You can separate your sample of soil in a similar way.

Fig. 12.17 Panning for gold. What is the principle of panning?

Experiment 12.21

Put about 5 cm of soil into a measuring cylinder. Fill the jar three-quarters full with water. Shake the cylinder to break up the soil, then set it aside for a few days.

Look at the surface of the water. Is there any plant material or humus floating there? Is the water clear or cloudy? How many layers of soil are present at the foot of the cylinder? Which layer will have the largest particles?

Experiment 12.22
Test your partner

Label three slides A, B, and C. Remove samples from the bottom layer of soil, from the top layer and from the water. Put one of the samples on A, but take care not to let your partner see which sample you put on A. Put another sample on B, and the third on C. Let your partner examine all three slides under the microscope. Ask him to tell you where each sample came from.

The fine particles 'hanging' or suspended in the water are **clay** particles. The upper layer of settled particles is probably **silt** or **fine sand**. Below them is **sand** and the largest particles are of **gravel**.

Fig. 12.18 A soil profile

12.14 COLLECTING SAMPLES OF SOIL

Fig. 12.18 shows a section through the soil. The soil is divided into two layers, the surface or top soil and the subsoil. Below the subsoil is rock. Here are possible methods that can be used to gather some soil.

(a) Dig up a spadeful of soil, then transfer it to a polythene bag.
(b) Use a soil auger, as in Fig. 12.19, to remove a long column of soil. Using this method you can sample both the top soil and the subsoil.

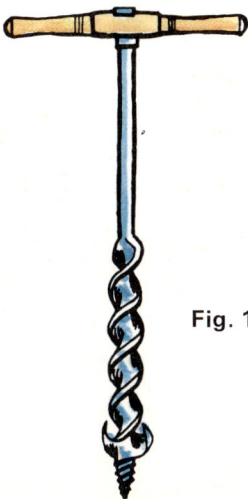

Fig. 12.19 A soil auger

(c) Can you devise a method of taking a sample of soil from the ground so that it is obtained exactly as it is in the ground? Discuss the suggestions of all the groups in the class and choose the most suitable.

Collect samples, using all three methods if possible, from as many different places as you can. Try to include soil from a woodland area, from a cultivated field, from ground near to a sandy beach, and from ground beside a river or from a hillside.

12.15 HOW MUCH WATER IS IN SOIL?

Experiment 12.23

For this experiment use soil collected by method (a). Divide the class into groups so that each group uses soil collected from a different source.

Weigh a lid from a large tin can. Add soil to the lid until there are 50 g of soil. Dry the soil thoroughly in an oven below 100 °C or on a warm radiator. Reweigh the lid and soil.
What weight of water has evaporated? What fraction of the original weight of soil is this? Compare your results with those obtained by other groups. What is the range of water in the soils?

Experiment 12.24

Now roast your sample of dry soil until it is red hot, then allow it to cool and weigh it. Heat it strongly again, then allow it to cool and reweigh. Is there a further decrease in weight? If there is you will have to roast the soil until you get no further loss in weight.
How much more weight has your sample lost? What is the total weight lost after roasting the dry soil? What did you see or smell from the soil when it was heated? What do you think was consumed in this roasting? Compare your results with those obtained by other groups.
When you roasted the soil you removed the humus. In which sample of soil would you expect the greatest amount of humus? Do your results confirm this?

Experiment 12.25

Powder your dry soil in a mortar. Using the sieves provided find the weights of different fractions of which your soil is composed. Enter your results as follows:

Weight of sample after removal of
water and humus   =   g

Weight of particles larger than
2 mm (gravel)   =   g

Weight of particles 0.5–2 mm (sand)   =   g
Weight of particles 0.1–0.5 mm
(fine sand)   =   g

Weight of particles less than 0.1 mm
(silt or clay)   =   g

Which is the main kind of particle in your sample of soil? Compare your results with those obtained by other groups.

Experiment 12.26
An additional experiment with humus

Try to reason out whether the amount of humus in the top soil is the same as the amount of humus in the subsoil taken from the same region. Collect a soil sample by method (b). i.e. using an auger, and test your hypothesis. Was your hypothesis correct?

12.16 HOW MUCH AIR IS IN SOIL?

Experiment 12.27

You have already devised a method to remove a sample of soil from the ground so that it is obtained exactly as it is in the ground (method (c)). Here is one possible way to do this (Fig. 12.20). Is it the same as your method? Collect a sample of soil by this method. Put 200 cm^3 of water into a large measuring cylinder. Empty the soil from the tin into the water. Shake the cylinder and allow the mixture to settle.
What is the volume of the mixture? Find the

Fig. 12.20 Collecting a sample of soil

1. Remove base from a can.

2. Lay can over soil, remove lid.

3. Push can into soil.

4. Dig away soil from one side of can.

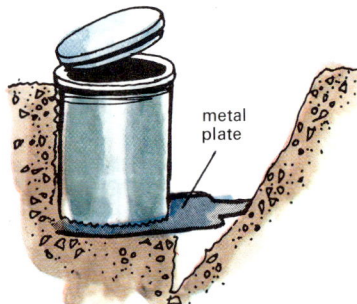

metal plate

5. Slide flat metal plate across base of can. Lift out can.

6. Tie polythene sheet to base of can.

Area from which soil was removed	% of water in soil	% of humus in soil	% of air in soil	Additional information
Woodland				
Cultivated field				
Hedgerow				

original volume of soil by measuring the volume of the tin. Add this volume to the volume of the water that was put into the cylinder, i.e. 200 cm³. What is the total volume? What volume did you actually get? How do you account for the difference? What fraction of your soil sample is air?

You can present your class results by preparing a large wall chart as shown above.

12.17 TYPES OF SOIL

A soil which is made up mainly of coarse particles is called a **sandy soil**. The air spaces between the particles are large and water drains through them quickly. Sandy soils are light and easily worked. A soil which is made up mainly of very fine particles is called a **clay soil**. The air spaces are very small and as a result water does not pass through them quickly. The soil becomes heavy and waterlogged. A mixture of the two types of soil is called **loam**. Look at your class results. Do you have sandy or clay soils? If so where did each kind of soil come from? Which kind of soil is most suitable for growing plants? Give reasons for your choice.

12.18 MAN'S EFFECT ON THE SOIL

Growing plants remove mineral salts and nitrogenous substances from the soil. These are built up into the tissues of the plants. When leaves fall from the trees and rot, and when plants die and decay, mineral salts and nitrogenous substances are returned to the soil. In this way vital sub-

stances pass from the soil to the plants and back to the soil again. But what happens if the plant material is picked and then removed? Look back at Experiment 12.24 and compare the amount of humus in woodland soil with the amount in soil taken from a cultivated field. There is less humus in the soil taken from the field. In this case fewer mineral salts and nitrogenous substances are

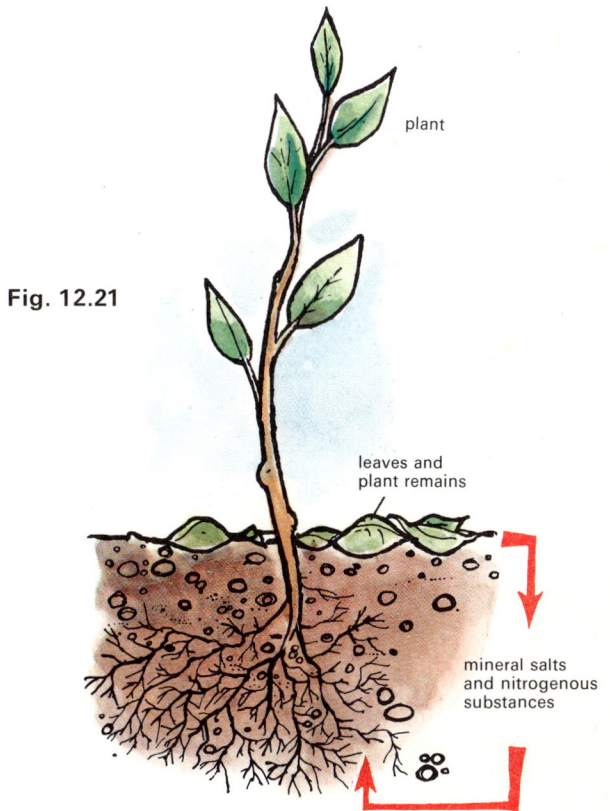

Fig. 12.21

plant

leaves and plant remains

mineral salts and nitrogenous substances

being returned to the soil. If the same crop is grown every year in the same field and harvested, it will remove so many valuable substances that the soil will eventually be unable to support further growth; in this way the natural balance of the soil is disturbed. However, some of the substances removed by plants can be put back into the soil by rotating the crops grown in one field or by adding fertilizers.

12.19 CROP ROTATION

Cereal crops remove much of the nitrogen compounds from the soil. Bacteria found in the roots of such crops as clover and peas manufacture nitrogen compounds. If these crops are grown alternately with cereals in the same field, the nitrogen compounds in the soil should not disappear. Here is a four-year rotation.

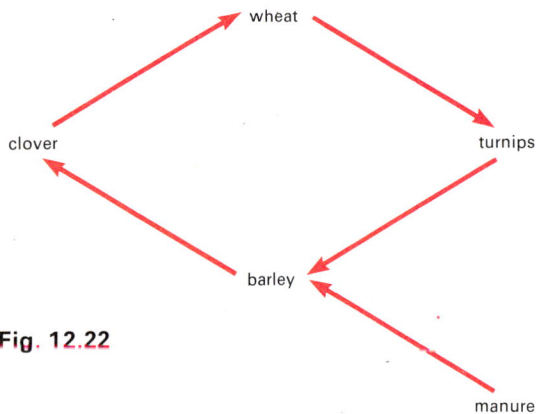

Fig. 12.22

Why was manure added before the barley was planted?

In order to provide ground for growing crops, man cleared vast areas of woodland. The ground was used extensively for growing one kind of crop, and when the harvests became poor the land was left and new ground cleared and planted. The deserted soil had few plants left in it to bind it together and so the top soil was blown away by the wind, or washed away by the rain and rivers. Vast areas of land became **eroded** in this way. Because of man's influence, areas that were once fertile became desert regions.

Today there is a great shortage of food and man is trying to turn vast areas of desert into fertile land suitable for growing crops. Find out where this is taking place on a large scale and what methods are being used.

12.20 CONSERVATION

We are now having to spend a great deal of time and money to correct mistakes made by man in the past. Because of the demands for food we must make use of all available lands, but we must use them sensibly and not destroy them. We must use the land for ourselves but we must also leave it so that it can still be used many years from now.

12.21 ANIMALS IN THE SOIL

Have you ever thought what it would be like to live in the soil, not in a building under the soil, but actually in the soil? It is dark, cold, and damp. This does not sound very suitable for us, but soil is the habitat for enormous numbers of animals. What does the soil provide for all these animals? You know that it contains mineral particles, water, humus, and air. Can you think how these substances are used by animals? List the four substances and opposite each write down a possible use. Discuss your ideas amongst the class and see how many uses you can get for each substance.

Experiment 12.28
Animals in the soil

Collect soil from the areas chosen for Experiment 12.23. Spread the sample on to a sheet of paper or an enamel tray. Pick out any animals you see. Using the key below and reference books, try to name the animals. Compare the animals from different soils.

Key for soil animals

1.	Body segmented	2
	Body not segmented	11
2.	Body without legs	3
	Body with legs	4
3.	Skin slimy, pinkish brown in colour	Earthworm
	Skin hard, light brown in colour	Leatherjacket
4.	Body divided into two distinct segments, four pairs of legs	5
	Body divided into more than two segments	6
5.	Body about 0.5 mm in length	Mite
	Body more than 0.5 mm in length	Spider

6. Body with three pairs of legs   7
   Body with more than three
   pairs of legs   9
7. 'Pincers' at end of abdomen   Earwig
   No 'pincers' at end of abdomen   8
8. Body 1 mm long, forked
   spring at end of abdomen   Springtail
   Body more than 1 cm long,
   hard shiny covering   Beetle
9. Body oval   Woodlouse
   Body long   10
10. One pair of legs per segment   Centipede
    Two pairs of legs per seg-
    ment   Millipede
11. Minute body, thread like   Nematode
    Body slimy, grey or black in
    colour   12
12. Shell present   Snail
    No shell   Slug

snail (5 cm)

woodlouse (3 cm)

centipede (3 cm)

Fig. 12.23

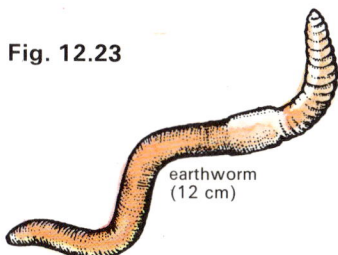

earthworm
(12 cm)

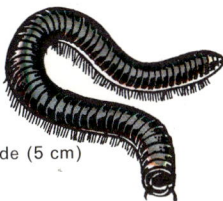

millepede (5 cm)

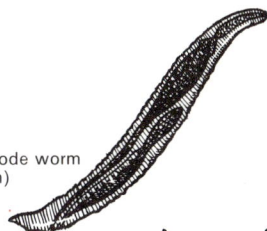

nematode worm
(1 mm)

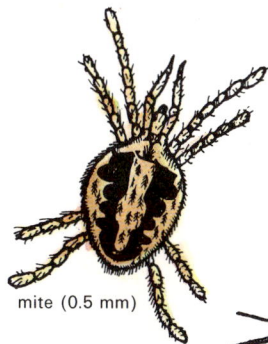

mite (0.5 mm)

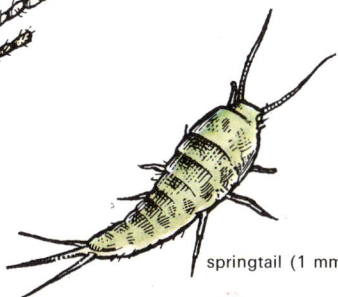

springtail (1 mm)

earwig
(2 cm)

beetle (2 cm)

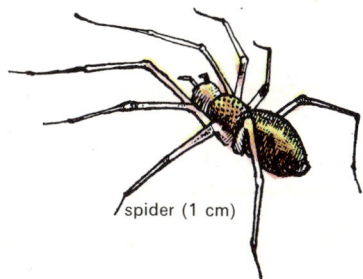

spider (1 cm)

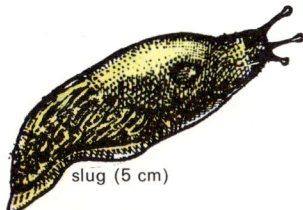

slug (5 cm)

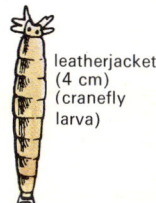

leatherjacket
(4 cm)
(cranefly
larva)

Experiment 12.29

Look again at the soil on the tray. Have you removed all the animals from it or are there still some hiding between the particles? If there are any we can make them move from the soil particles by changing the conditions of the soil. If we heat up the damp cool soil, it will become dry and warm, and the animals will move out of it.

bench lamp
(source of heat and light)

large plastic funnel

soil

wire gauze

alcohol or formalin

Fig. 12.24

Fig. 12.24 shows a Tullgren funnel. Place a sample of soil on the wire gauze in the funnel. The light above the soil is the source of heat. As the surface of the soil becomes warm and dry, the animals in that layer will move downwards and finally fall into the collecting jar outside the funnel. Leave the funnel set up for several hours, collect any organisms from the collecting jar and try to identify them.

Experiment 12.30

Make a guess about the number of animals in a sample of top soil from a woodland compared with the numbers in a sample of subsoil from the same area. Using the sampling methods in Experiment 12.28 and Experiment 12.29, test your guess.

Experiment 12.31

light

glass rod

water

muslin bag + soil

glass filter funnel

rubber tubing

clip

collecting jar

Fig. 12.25

We can use the apparatus in Fig. 12.25 to extract microscopic animals which live in soil water. Set up a Baermann funnel and leave it for several hours. Take a drop of liquid from the collecting tube and examine it under the microscope. Draw any organisms you see.

The organisms we have been finding in soil range in size from earthworms which may be 12 cm in length to minute animals which are only a fraction of a millimeter. Have you found the smallest organisms using a Baermann funnel or are there any even smaller?

Experiment 12.32

Pass a sample of soil through a fine sieve. Divide the sample into two. Put one sample, A, into a crucible and roast it. Allow it to cool, then empty the contents of the crucible into a muslin bag. Put the second sample, B, directly into a muslin bag. Suspend each bag above bicarbonate indicator in a sealed flask. Leave the flasks for one day, then examine the indicator in each.

What has happened to the indicator in A? What has happened to the indicator in B? What gas is present in flask A? What gas is present in flask B? Where have these gases come from? What can you deduce about the two samples of soil? What changes have occurred in soil A, as a result of roasting?

Carbon dioxide was liberated from the sample of fresh soil. The liberation of this gas can be used as a method to detect living organisms. All living things respire and give out carbon dioxide. There must be living organisms present in soil which are too small to be seen. In this case these micro-organisms are *bacteria*.

12.22 BACTERIA

Bacteria are present in soil. We cannot see a bacterium because it is too small. However if we grow them, we can get them in numbers sufficient to see them. To do this we must provide them with food.

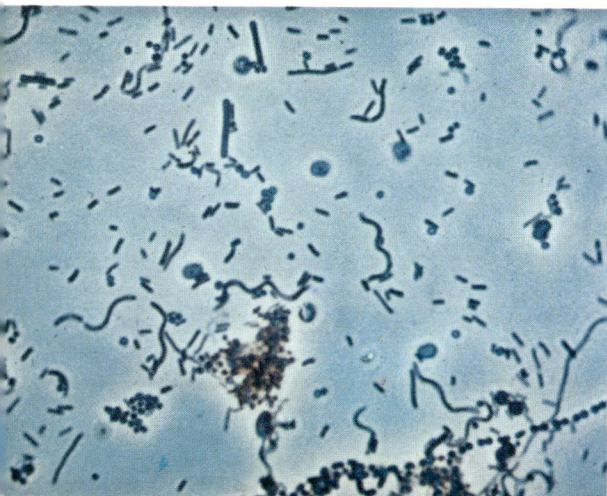

Fig. 12.26 A photomicrograph of mixed bacteria of different shapes

Experiment 12.33

Your teacher will give you a flat, transparent dish called a Petri dish, containing a clear jelly. This jelly contains a food extract. The dish and jelly have been sterilized, i.e. heated under pressure to kill any bacteria in them.

Shake up a sample of fresh soil in boiled water. Place two drops of this liquid on to the jelly in the Petri dish. Replace the lid and then do not remove it again from the base of the dish.

Shake up some roasted soil in boiled water. Add two drops of this to a second Petri dish, and cover it as before. Label each dish and leave them in a warm place or in an incubator set at 32 °C for a few days. Bacteria grow quickly under warm conditions.

Observe and draw what you see after the incubation period. The creamy-coloured circles on the jelly are colonies of bacteria. DO NOT REMOVE THE LID FROM THE DISH. Some bacteria can be harmful to us and cause disease, so you must never touch any contaminated jelly.

Experiment 12.34
Can soil bacteria grow in any other food?

Set up experiments shown in Fig. 12.27 Leave the flasks in a warm place for 24 to 48 hours. Remove the cotton wool and note the smell in each. Note the appearance of the milk in each. Why is the milk different in the two flasks? Why were the flasks bunged with cotton wool?

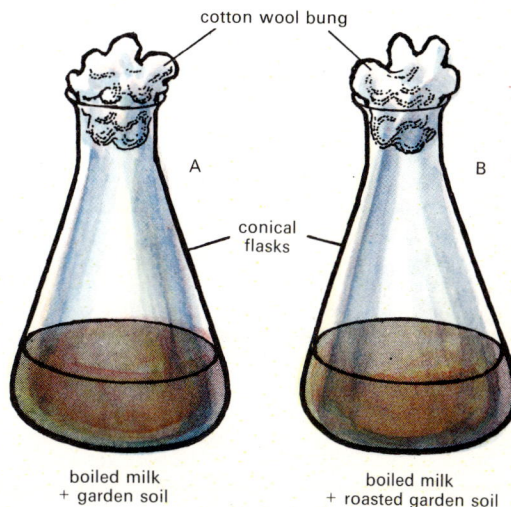

Fig. 12.27

Experiment 12.35
Do other substances contain bacteria?

Label five sterile Petri dishes containing sterile jelly, A, B, C, D, and E. Treat them as follows:
   To A add two drops of tap water.
   To B add two drops of milk.
   To C add two drops of pond water.
Scrape dirt from under your fingernails. Add this to boiled water. Add two drops of this liquid to the jelly in plate D. Leave E untreated. Leave the plates in a warm place for a few days. Observe and draw what you see. Where do bacteria occur? Why did you set up plate E? What do you call this plate?

12.23 OTHER MICRO-ORGANISMS

Experiment 12.36

Soak a piece of bread in water. Remove it and squeeze gently to remove excess water. Brush some dust up from the floor and put this on top of the bread. Place the contaminated bread in a plastic box and leave it in a dark place for a few days. Describe the appearance of the bread.
   Scrape some of the felt-like material from the bread and examine it under the microscope. Draw what you see.

   The thread-like structures you saw under the microscope are part of one of an unusual group of plants called **fungi**. Leave the bread for a further two weeks and examine it every time you are in the laboratory. What changes in colour occur with standing? Is there one or more fungi present? How much of the original bread remains? Explain these changes in the bread.

Fig. 12.28

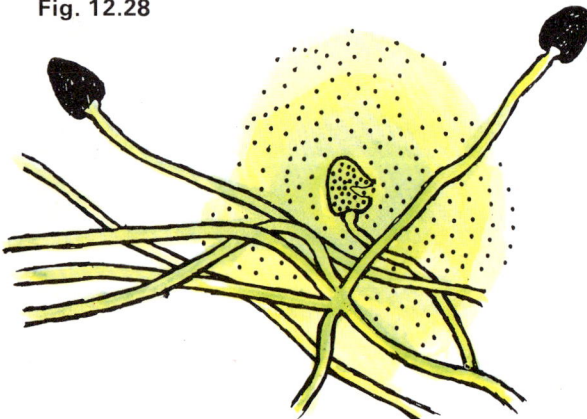

Experiment 12.37
An additional experiment

Plan an experiment to find the optimum (i.e. the best) conditions for the growth of fungi. The conditions you could test are (a) temperature, (b) amount of water, (c) source of food.

   'Damping off' is a fungus disease of plant seedlings. If plant seedlings are grown close together and overwatered they may become contaminated with the disease.

Experiment 12.38

Conduct experiments to grow cress seedlings (a) free from the disease, (b) with the disease. Remove a contaminated seedling and examine a small part of the stem under the microscope. What do you see?

12.24 USEFUL MICRO-ORGANISMS

Experiment 12.39

Mount a small quantity of yeast in water on a slide and examine it under the microscope. Draw what you see.

   Yeast is a fungus which is used by man in the production of bread and alcohol. Let us see what properties this fungus has which make it so important.

Experiment 12.40

Mix two samples of bread dough. Add yeast to only one of them. Allow them to 'prove' and then bake each sample. What do you observe about the samples?

Experiment 12.41

Label three test-tubes A, B, and C.
   To A add sugar solution and yeast.
   To B add sugar solution.
   To C add yeast suspension.

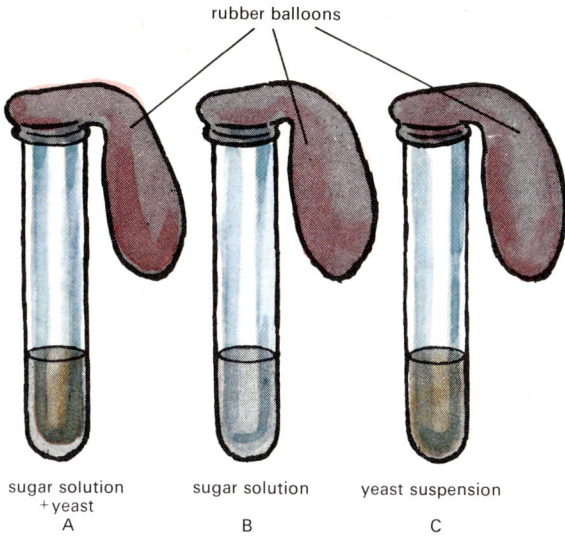

rubber balloons

sugar solution
+yeast
A

sugar solution

B

yeast suspension

C

Fig. 12.29

Cover the mouth of each tube with a balloon. Place the tubes in a water bath at 37 °C for about half an hour. What do you observe? What has been produced in tube A? What do the control tubes indicate?

Tightly grip the neck of the balloon on tube A. Remove the balloon. Taking care not to let any gas escape, hold the neck of the balloon below some bicarbonate indicator. Squeeze the gas into the indicator. What do you observe?

What gas has been liberated from the contents of tube A? Where has this gas come from? What function do you think the yeast has in baking bread?

Stopper tube A. Leave it in a warm place in the laboratory for about one week. After this time remove the stopper and smell the contents of the tube. What substance is now present in the tube? By what process could you separate this substance from the sugar solution and yeast?

Fig. 12.30

balloon

bubbles of gas

bicarbonate indicator

Fig. 12.31 A variety of fungi

12.25 OTHER IMPORTANT FUNGI

A fungus which does a great deal of damage to old buildings is 'dry rot'! It feeds on and weakens wood, and can spread through stone from one wooden floor to another. Although it is called dry rot, it starts growing in wood that has been wet.

Penicillium is also a fungus but this one is very useful to us.

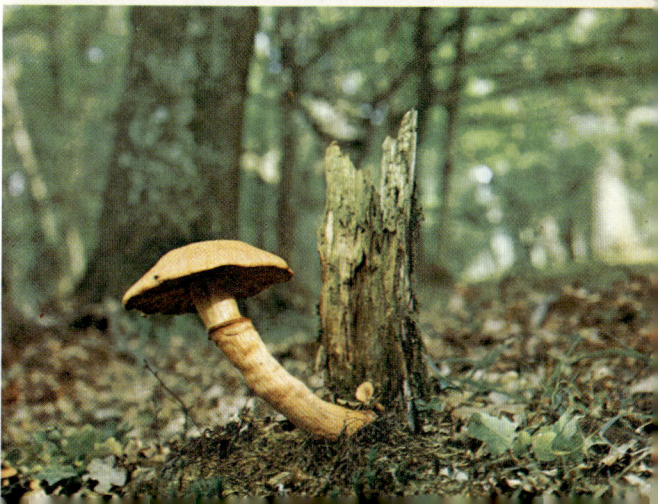

Experiment 12.42

Your teacher will give you a Petri dish in which bacteria have been grown. How can you tell that there are bacteria on the jelly? Using sterile forceps, place a disc containing penicillium on to the bacteria growing on the jelly. Label the dish and incubate it at 32 °C for one to two days. Look at the area around the disc. What do you see? What does this indicate?

The drug penicillin is extracted from penicillium. It kills bacteria and so it is used as an antibiotic. Using books in your library, find out about the discovery of penicillin by Sir Alexander Fleming.

WHAT YOU HAVE LEARNT IN THIS UNIT

1. The earth has a radius of about 6000 km. It is made up of an outer crust, only a few kilometres thick, then a semi-plastic mantle, followed by a liquid core and then a solid inner core. Fig. 12.7 on page 43 illustrates this.

2. Some rocks have solidified from molten material. These are called **igneous** rocks. They are crystalline and contain no fossils.

3. Rocks which have been deposited under water are called **sedimentary** rocks.

4. Rocks which have been formed from others by the action of heat and pressure are called **metamorphic** rocks.

5. Not many metallic elements are found as such in nature. This is because metals often combine with oxygen and sulphur, both of which occur abundantly.

6. Minerals from which metals are obtained are called ores. To obtain metals, ores are smelted. When they are heated they are usually converted into oxides. Then the oxygen has to be pulled away from the metal by heating it with a reducing agent such as carbon, which is used in the form of coke.

7. When most carbonates are heated, they give off carbon dioxide and leave the oxide of the metal. Exceptions are sodium and potassium carbonates which cannot be broken down by heating them.

8. Coal is the fossilized remains of vegetation. There are various kinds of coal depending on its age. They differ in the ease with which they burn. We have peat, lignite, bituminous coal, and anthracite. By heating coal out of contact with the air, a gas (coal gas), tar, and coke are obtained. All these are very widely used in industry. Many useful products can be obtained from coal-tar by distillation.

9. Oil has been produced by the decay of sea animals and plants. As obtained from the oil wells it is a thick liquid, and is a mixture of various kinds of oil. These are separated from it by fractional distillation. It yields petrol, paraffin (or kerosene), lubricating oil, Vaseline, wax, and pitch.

10. Sea water contains many dissolved salts, the largest in quantity being sodium chloride (common salt). All sodium chloride is obtained either from the sea itself or from salt deposits which came from ancient seas now dried up.

11. Soil is made up of tiny particles which have been broken off rocks by the wind, rain, frost, and other agents, together with decayed organic matter called humus. There are three main types of soil – clay soil, sandy soil, and loam (a mixture of clay and sandy soils).

12. Many small animals live in the soil. Some are so small that they can be seen only with a very high-powered microscope. Some of these very small organisms are called bacteria.

13. Other micro-organisms you have met are fungi and moulds. Yeast and penicillium (from which penicillin is made) are very important fungi.

Unit Thirteen
Support and Movement

13.1 THE IDEA OF FORCE

In this unit we are first going to find out something about forces. A force is not something we can see, but it is something we can all give and receive. Through which of the senses do we know about forces? From our sense of touch and the feeling of discomfort in our muscles we are made aware of when we are exerting a force on anything. We might be lifting a heavy object, pushing or pulling a garden roller or a lawn mower, or twisting the cap off a lemonade bottle. In all these cases we exert forces, and we are aware of the fact because our muscles get tired. Can you think of other examples of forces? Sometimes,

Fig. 13.1 Exerting forces

no matter how hard we try, we are unable to exert a sufficiently large force to move a heavy object. We could not, for instance, lift a car by our own unaided efforts. Write down the names of some of man's inventions which enable him to do jobs like this.

Fig. 13.2 Modern lifting gear for moving very heavy objects

13.2 TYPES OF FORCE

To get an idea of the different types of force you will be given some objects on which you can use forces. You can use your hands, or you can hang or place weights on these bodies and see what happens. The objects should include a lump of Plasticine, some springs, elastic bands, foam rubber, a copper wire, and a steel wire. When you are doing these experiments try to keep in mind the picture we built up in Unit 4 of what solids are like. You will remember that in some solids which we called crystals, the particles were arranged in ranks and files, and that there are spaces between them. There must be some kind of force keeping the particles apart, but at the same time preventing them from being pulled apart easily. You will remember the effect of rubbing cellulose acetate and polystyrene strips and how they attracted each other. Scientists think that similar electrical forces provide the attraction between the particles in a solid and hold the material together.

13.3 THE EFFECT OF FORCES ON THE SHAPES OF BODIES

Experiment 13.1
Pushing

Try the effect of pushing between your hands a solid rubber ball, or an eraser, a piece of foam rubber, and a bed spring. You are trying to compress the material into a smaller space. Do you think you are really succeeding? If we stop trying to compress the material, what happens? We say that an object that goes back to its original shape and size when we stop trying to compress it by pushing is 'elastic'. Try doing the same thing with Plasticine. Is it elastic?

Fig. 13.3

Now twist, in turn, the block of rubber and the Plasticine. What happens when you stop this time? In the case of the rubber the forces between the particles must be very strong to bring them back to their original shape. The force between the particles in the Plasticine must be rather different. In what way, do you think?

Experiment 13.2
Stretching

Hang a heavy weight on first a steel wire and then a copper wire of the same diameter. What happens in each case? In which are the atoms held together by stronger forces? Why should the wires have the same diameters? If you use the right weights you will find that the steel wire stretches, while the copper wire stretches more and finally breaks.

Experiment 13.3
Pulling

Try pulling the Plasticine and rubber blocks between your hands. In which material are the forces between the molecules weaker? Does this agree with what you found in Experiment 13.1?

Do you think that a material like rubber or Plasticine would be suitable to make our limbs or those of animals? Perhaps our muscles and tendons which join on to the bones seem rather elastic, but this property would hardly be a suitable one for the material of the bones themselves.

13.4 THE FORCE OF GRAVITY

Experiment 13.4

Roll a marble along the bench. What happens to it when it reaches the edge? It does not matter what the marble is made of; it can be glass, porcelain, or steel, and exactly the same kind of thing will happen.

Why does the marble fall down? There must obviously be some force acting on it that makes it do this. Do you think it is the weight of the air above us that makes the marble fall? Which do you think would fall faster and reach the ground quicker if dropped from the same height – a marble or a golf ball?

We can get an answer to both of these questions by trying another experiment.

Experiment 13.5

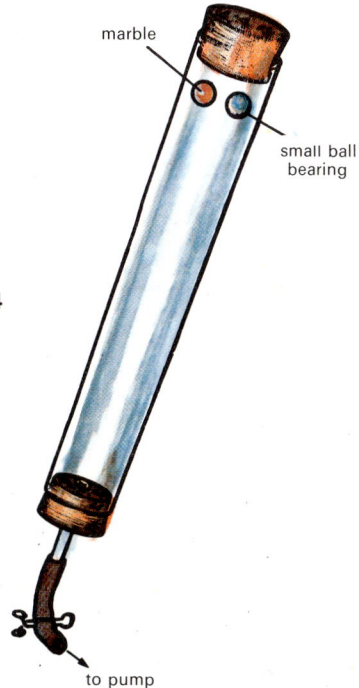

marble

small ball bearing

Fig. 13.4

to pump

Put a marble and a small ball bearing in a glass tube like that in Fig. 13.4 and quickly invert it. Which reaches the end of the tube first? Now pump the air out of the tube and repeat the experiment. Is there any difference? Repeat the experiment with a small coin and a paper disc the same size as the coin instead of the balls.

When the air is removed from the tube, the coin and the paper fall, and surprisingly they reach the bottom of the tube at the same time, as did the two balls.

When there was air in the tube the paper disc fell more slowly than the coin. Can you think of an explanation for this? Perhaps the air was rubbing against the light piece of paper and the friction force produced partly balanced out the effect of the force of gravity.

This experiment was first done with a coin called a 'guinea' and a feather. Would these articles give a different result from yours?

13.5 THE EFFECT OF A FORCE ON MOTION

Experiment 13.6

Most of you will have had the experience of pushing a heavy wheelbarrow or roller. To keep the object moving along a flat path you have to keep on pushing all the time. See if

pull!

Fig. 13.5

push!

there is any difference in that situation and when you start a roller skate, or what we call a 'dynamics cart' rolling along the bench top or along the smooth flat floor of the classroom or the corridor. Do you need to keep on pushing this time to keep the roller skate moving? Try rolling a large marble or ball bearing on the floor. How far does it go before it stops?

As with the roller skate so with the ball bearing. These objects do not 'want' to stop once they have been set in motion. Have you noticed this effect with your cycles? When cycling along a flat, smooth road it is sometimes possible to 'free-wheel' for a long time before the cycle slows down. What would cause your cycle to stop if you are free-wheeling? You can no doubt think of many reasons. The most common are:

1. putting on your brakes, and making the brake blocks rub on the rim of the wheels;

front forks of cycle

Fig. 13.6 Where does friction act here?

2. the fact that you have not oiled your cycle well enough, and there is rubbing at the axle; and

3. the wind pushing against you.

In the first two cases, the rubbing has produced what we call a force of **friction**. You will remember that in Unit 3 we saw that this could be due to roughnesses in the surfaces locking into each other.

In the case of the objects we have just been considering – cycles free-wheeling, skates and balls rolling, we have found that once these objects are made to start moving they keep on moving by themselves and they will only be stopped when another force, such as friction, pushes against them. Why does a wheelbarrow or a roller have to be pushed all the time to make it move? What force, do you think, is trying to stop these implements from moving?

Later on you will try pulling a roller skate along with a steady force and you will find that the speed with which it moves is not steady. What happens is very like when a ball falls from a high building. As it falls towards the earth it moves faster and faster. What force is acting on the ball all the time? Is it a steady force?

We find that when there is no friction or other opposing force, a cycle, a ball bearing, a roller skate, or anything else when set in motion by a force will continue moving and not stop. Of course, if there were no force acting to set these things moving, they would simply stay still. When there is an extra, or unbalanced force, such as the extra pull on the roller skate along the bench, or the pull of gravity on the falling ball, the motion changes, and gets faster and faster. We call this **acceleration**. The accelerator in a car makes it go faster and faster because it causes more petrol to be burnt in the engine. This makes the engine exert a bigger force, so the car moves faster.

13.6   MOTION WITHOUT FRICTION

Experiment 13.7

If possible set up a 'linear air track'. This is a long metal tube with many holes drilled in its sides. Air is blown into the tube from a vacuum cleaner unit working in reverse, so that air puffs out of the holes. A 'vehicle' fits over the tube and can move along it. There are rubber buffers at the ends of the tube at which the vehicle rebounds. When the vehicle is set travelling on the tube, does it seem to slow up quickly? Can you suggest why it seems to continue its motion for so long? What force have the jets of air from the track got rid of, or 'balanced out'?

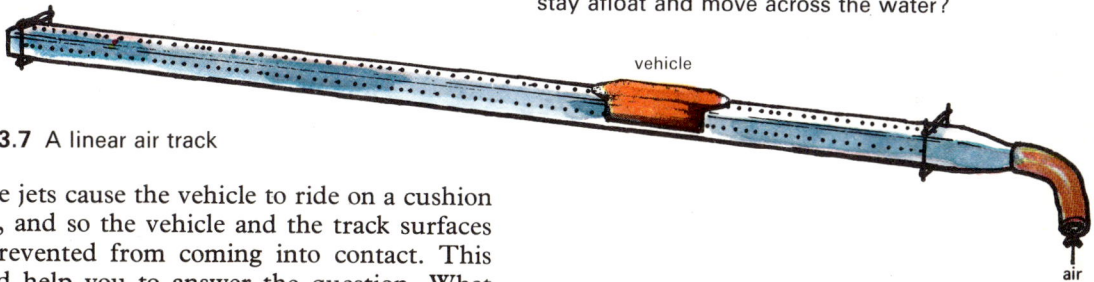

Fig. 13.8 Hovercraft on land and sea. How does it stay afloat and move across the water?

Fig. 13.7 A linear air track

The jets cause the vehicle to ride on a cushion of air, and so the vehicle and the track surfaces are prevented from coming into contact. This should help you to answer the question. What modern means of transport uses this idea? Some of you may have travelled on, or seen films of, a hovercraft. Find out how it works.

In ice hockey there is hardly any friction at the surface of the ice to slow the puck down.

Experiment 13.8

You can make a balloon puck by following the instructions in Fig. 13.9. Blow up the balloon and hold the neck tightly so that the air does

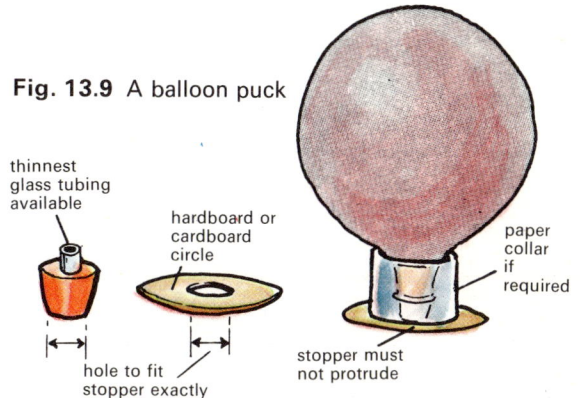

Fig. 13.9 A balloon puck

thinnest glass tubing available

hardboard or cardboard circle

paper collar if required

stopper must not protrude

hole to fit stopper exactly

not escape. Fit the neck over the stopper, set the puck on the bench, and give the puck a push. If the bench is smooth and polished, the puck should travel the length of the bench quite easily.

Perhaps, when you have successfully made your 'balloon puck' and seen it coast along at a steady speed once you start it off, you can more readily understand one aspect of the motion of a space ship in a circular orbit round the earth with its rocket motors switched off. It is able to do this if it has reached a speed of about 28 000 kilometres per hour (18 000 miles per hour) and provided it is just outside the atmosphere where there is no air and so no friction to slow it down.

13.7 MEASURING FORCE

In many experiments it is very important to be able to measure accurately the size of a force. This can be done by using a spring balance. What happens to the length of spring when it is pulled? We can use this increase in length of a spring to measure the size of the force stretching it.

Experiment 13.9
Stretching a spring

Set up the apparatus in Fig. 13.10 and collect a number of identical objects (e.g. a set of washers all the same, a set of cotton reels all the same, a set of slotted weights, etc.).

Choose your objects so that they will produce a reasonable stretching of the spring when they are hung on to it. Mark the position of the pointer attached to the hook on the spring without any objects being added. Now add one object and again mark the position of the pointer. Repeat this process several times, adding one more object each time. What do you notice about the extra stretch, called **extension**, which each of the identical objects produces? Make a graph of extension against force. What kind of graph do you get?

Each time an extra weight is added you have found that there is the same extension. Because each object is identical they are attracted to the earth with the same force, so what we have found is that equal forces produced equal extensions of the spring. This was discovered over 300 years ago by Robert Hooke and is called Hooke's Law.

You have made a scale on the card and should be able to use your spring to find the weight of suitable objects in terms of the units of weight you have used. Thus you can find the weight of a ten pence piece in terms of 'washer weights'.

13.8 THE UNIT OF FORCE; THE NEWTON

You and your classmates may have used quite different sets of objects in making the scale on your spring balance. This would be very inconvenient in practice, so we have to decide on one particular unit of force which all scientists will use. This unit of force is called the newton. To give you some idea of the size of a newton we may say that it is about the weight of a medium-sized dessert apple. The unit is named, of course,

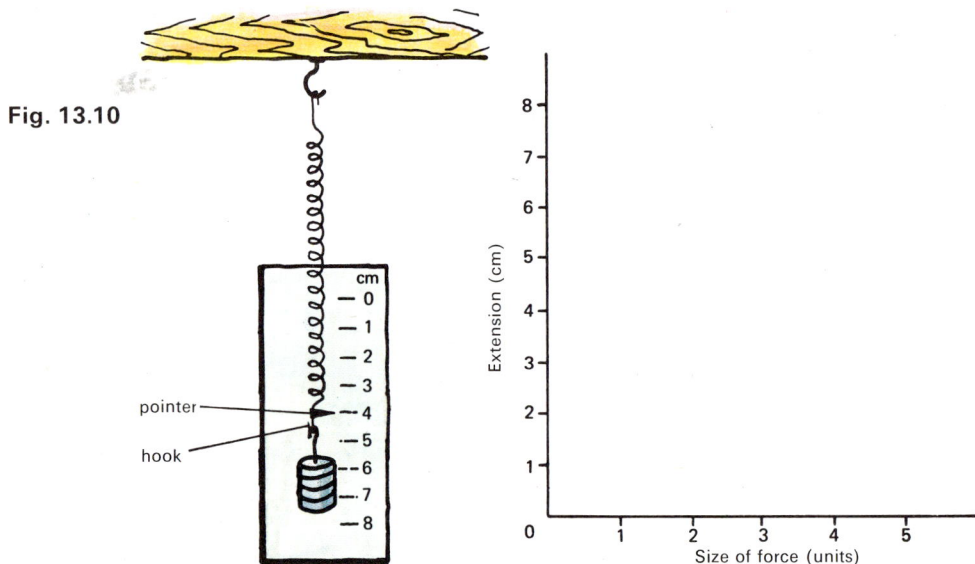

Fig. 13.10

after the great scientist Sir Isaac Newton. He is famous for his work on the study of gravity, and he is supposed to have had his interest aroused in this topic by watching apples fall to the ground in an orchard. Find out all you can about him from a book in the library.

Experiment 13.10
The newton spring balance

In your laboratory you will have some spring balances calibrated in newtons. Using one of them, find out the weights of some masses in newtons. You should find out that the weight of 0.1 kg (100 g) is just about 1 newton. This is written 1 N.

Fig. 13.11

newton
spring
balance

apple

Some additional experiments

1. Add gramme masses in turn to the spring you calibrated and find out how many grammes one of your units represents.
2. Find out how many newtons one of your units represents.
3. You should have found that the graph of the extension of the spring against force was a straight line. Add more and more weights to your spring. What shape of graph do you get now? Would your spring be suitable for measuring large forces?
4. If you did not make up the 'straw balance' described in Book 1, page 8, now is a good time to construct one. What range of weights is it specially suitable for?

13.9 THE LEVER

Experiment 13.11

Fit up the apparatus shown in Fig. 13.12. Carefully adjust the position of the wooden block and elastic band so that the wooden lever is balanced. There are a number of experiments we can do to balance the lever with different forces, but some of the more important ones are suggested in the incomplete table below. Fill up the spaces where there are blanks. In Group 1 the weight at B is a single slotted weight, and at A there are one or more slotted weights piled on top of each other. In measuring *x* and *y*, the distances of the weights from the elastic band, remember that the middle of the weight has to be exactly on the line on the scale.

		Force at A	Distance *x* (cm)	Force at B	Distance *y* (cm)
Group 1	(a)	1 unit	20	1 unit	?
	(b)	2 units	?	1 unit	20
	(c)	4 units	?	1 unit	20
Group 2	(a)	2 units	12	3 units	?
	(b)	4 units	9	3 units	?
	(c)	3 units	10	2 units	?

In Group 1 you should find that one of your units of weight at A kept at the same distance from the balance point (called the pivot or fulcrum) can balance different weights placed at B, different distances from the pivot. In case (c) the single unit will be able to balance the effect of a weight four times its own size, but do you notice that this single weight has to be four times further away from the pivot to do this? Here we see one of the important benefits of a lever – it can magnify the effect of a small force.

In the results of Group 2 do you notice any mathematical connection between the figures in the left-hand two columns and those in the right-hand columns?

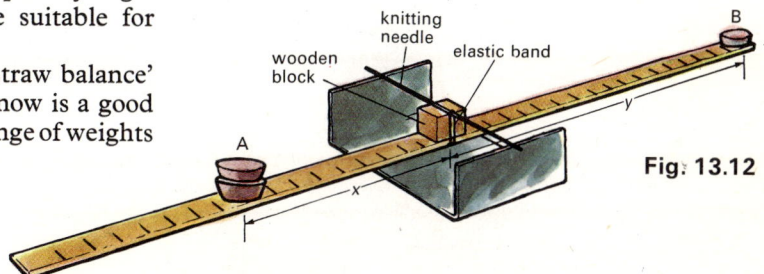

knitting
needle
elastic band
wooden
block
A
B
y
x

Fig. 13.12

Levers are used very widely in everyday life. Generally we make use of the fact that a small force has more leverage or turning effect the further it is from the pivot.

Fig. 13.13

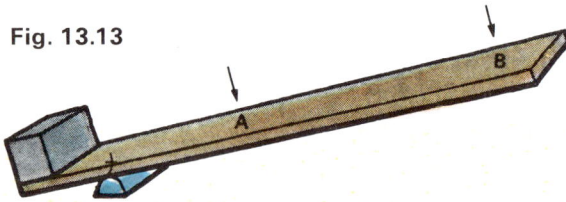

Look at Fig. 13.13. Where would you choose to try to lift the stone – by pushing down at A or at B?

Something more for you to do

1. Try shutting a heavy door by pushing first near the hinge (pivot), and then at its edge. Why do you think door handles are placed where they are?

2. Compare the ease of cutting with scissors near the rivet and near the end of the blades.

3. Find where on the shafts of a loaded barrow it is easier to lift the barrow.

4. Use a claw hammer to pull nails out of a piece of wood grasping the handle near the head, and then at the end of the shaft. In which case is the work easier?

13.10 PAIRS OF FORCES

Here is a group of experiments of which you should try as many as possible. It does not matter in what order you do them.

Experiment 13.12
Pairs of forces

1. You and your partner should put on roller skates and face each other. Both push at the same time. Who moves? If *you* do the pushing and not your partner, who moves this time?

Fig. 13.14

Fig. 13.15 Levers in sport

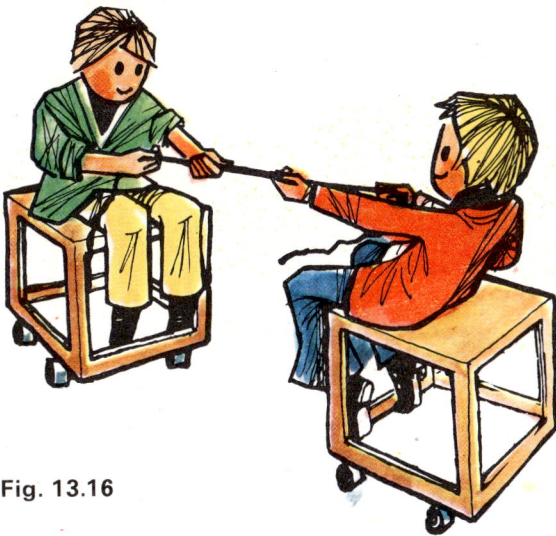

Fig. 13.16

5. Fill a water rocket half full with water and then pump in some air with about twenty strokes of the pump. Now release the rocket. This experiment is best done in the open air. It is a good idea to wear waterproof cuffs.

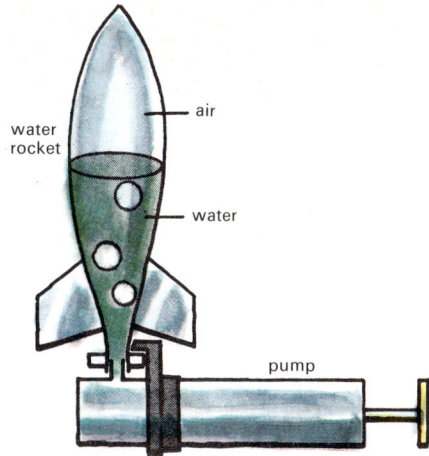

Fig. 13.19

2. This time you should each be seated on a science department trolley about 2 or 3 metres apart. Each should hold an end of a piece of stout cord. What happens when you *both* pull the cord? What happens when only *you* do the pulling?

3. Compress the spring of a dynamics cart or trolley, and place it on its own on the bench. Now trigger off the spring. Does the cart move? Now compress the spring again, place the cart end to end with another cart, and trigger off the spring. What happens this time?

Fig. 13.17

4. Blow up a long thin balloon and hold the neck of the balloon tightly so that the air cannot escape. What happens when you release the balloon? Another way to do this experiment is to use a balloon attached to drinking straws threaded on to thin string stretched across the room (Fig. 13.18).

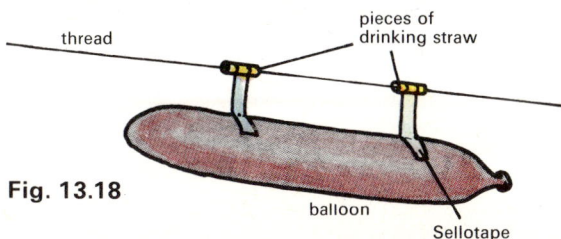

Fig. 13.18

6. Scatter a thin layer of polystyrene beads on a large tray and cover this with a piece of paper. Now 'wind up' a friction-drive toy car and with its wheels spinning put the car on top of the paper. What happens?

Fig. 13.20

There are many important things to discover in this experiment. In part 1 you found that when you pushed your partner he was not the only one to move; you also moved in the opposite direction. So in this case two forces acting in opposite directions were involved.

In part 2, when you pulled on the string both trolleys moved towards each other, so two forces must have been acting. You pulled on your partner and, because of this, he was pulling on you too, although he may not have been consciously intending to do so.

Fig. 13.21 A multiple-exposure photograph of a rocket taking off

Triggering off the spring of the single dynamics trolley in part 3 did not result in the trolley moving off, but when the spring was able to push against another trolley, not only did the second one move off, but the first trolley was pushed in the opposite direction.

Parts 1 to 3 seem to indicate that forces occur in pairs in opposite directions, and this was also found in part 4. When the neck of the balloon was released, the air was forced out in one direction, while the balloon was pushed in the opposite direction. Similarly in the water rocket, where the water rushed out in one direction and the rocket took off in the opposite direction. Rockets which launch space-ships operate in this way too. Can you explain what moves in the opposite direction to the rocket in this case?

While jet and propeller driven aircraft can fly in the atmosphere, but not beyond it, rocket propelled space vehicles can move about and change direction beyond the atmosphere. Why is it that rockets can do this when conventional aircraft cannot?

In part 6 you found that when the toy car moves forward it pushes on the paper which moves backwards. What part did the polystyrene beads have to play in this?

13.11 WORK AND ENERGY

In Unit 3 (Book 1) we found out a great many important facts about energy. Just as forces are invisible, so also is energy. You will remember that we convert the chemical energy in the food we eat into mechanical energy. You would probably use most energy, and feel correspondingly tired, doing what we call manual jobs, where you might be lifting heavy loads, like coal or bricks. When you apply a force to lift up loads we say you are doing work. What kind of energy would the load have if you lifted it up to a higher level?

If you cast your mind back to Unit 3 you will remember that the chemical energy in your food was converted (at least partly) into kinetic energy as you moved the load up, and then into potential energy at the higher level. If you did a lot of very hard work of this kind what other form of energy would be made?

You will remember that we said that the unit of force is the newton, which is approximately the weight of a mass of 100 g (roughly a medium-sized apple). If you were to lift this mass through a height of 1 metre we say that you have done 1 newton-metre of work. The newton-metre is usually called a joule. Joule was a British scientist who did a lot of experiments on energy.

Copy the table below in your note book and try to complete it. In (b) if we require ten times the force as in (a), we will require to do ten times the amount of work or 10 J. Fill up each blank in turn in this way.

Object		Force required (weight 'N')	Height (m)	Work done = force × distance (N m or J)
(a)	apple	1 N	1 m	1 N m = 1 J
(b)	1 kg	10 N	1 m	?
(c)	2 kg	?	1 m	?
(d)	3 kg	?	1 m	?
(e)	3 kg	?	2 m	?

In part (e) you should calculate that 60 J of work would have to be done.

To get some idea of what work is you should lift various objects from the floor to the bench using a newton balance; you can calculate the work you have done in each case.

Experiment 13.13
Work done in dragging

Try dragging objects along the floor or the bench, measuring the force you have to use with a newton balance. Calculate the amount of work you have done in joules. Do you require to do less work to drag the object along than to lift it?

Object	Force required (from newton balance)	Distance (m)	Work done (N m or J)

Experiment 13.14
Finding your own power

Each time you climb the stairs at school or at home you are transforming your chemical energy into kinetic energy and then into potential energy. When you are doing work in climbing the stairs, what is the size of the force you are using?

If you know your own mass in kilogrammes (remember that 2.2 lb is equivalent to 1 kg) you could find out how much work you do in climbing the stairs. The amount of work you do is equal to the product of the force and the vertical distance through which it is raised. What other thing, then, besides your weight would you require to know if you wish to calculate the work you do? Find this out, and then calculate the work you do in climbing from one floor of your school to the next. Sometimes you might climb the stairs slowly, particularly if you are going to a rather unpopular subject, and sometimes you might run up the stairs if you are late for your class. Would the amount of work you do be the same?

The rate at which work is done is called **power**. It is measured in watts. You do not need to be told for what the Greenock-born man, James Watt, was famous. One watt is a rate of working of 1 joule in 1 second. It is good fun to find your power in watts. Find the time you take to run up the stairs, and divide the number of joules of work you have done by the time you took in seconds.

my mass = kg

$\Rightarrow$ my weight = N

vert. ht. of stairs = m

$\Rightarrow$ work I did = N m (J)

time I took = s

work I did in 1 s = J

$\Rightarrow$ my power = Js^{-1} (watts)

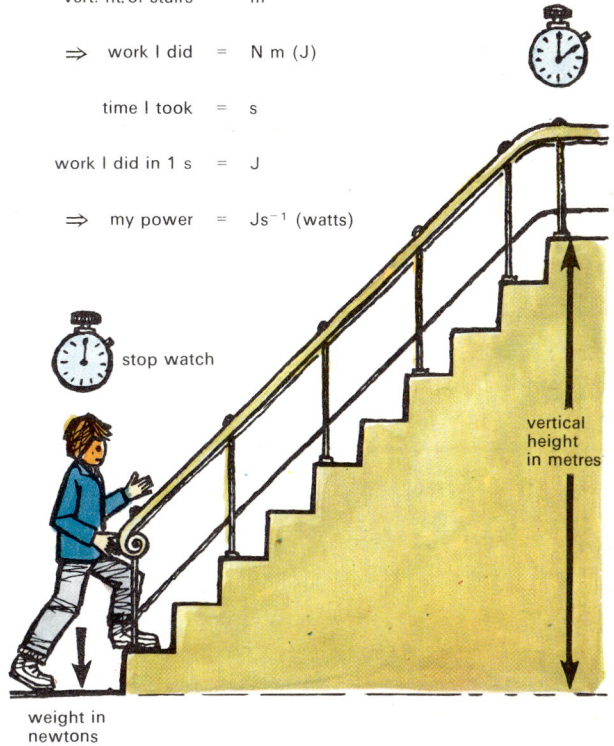

stop watch

weight in newtons

Fig. 13.22

13.12 MACHINES FOR LIFTING

Experiment 13.15

Take a long lever and hang a 2 kg load on it. For our purposes, we can say that this has a weight of 20 N. Now pull on the other end of the lever with a newton balance and find the

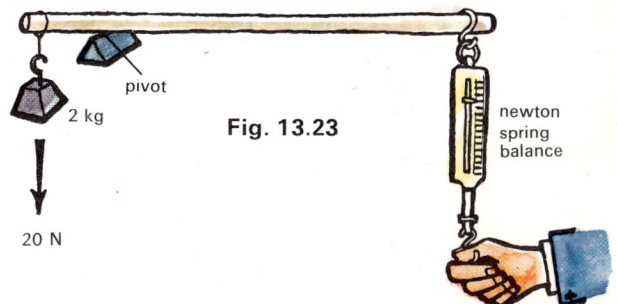

pivot

2 kg

20 N

Fig. 13.23

newton spring balance

force required to lift the load. This force is called the effort. Which is greater, the effort or the weight of the load?

When you move the load through a height of 5 cm how far do your fingers, pulling on the spring balance, move? Calculate the amount of work done as the load moves, and also that done by you in pulling on the balance. Notice that you have done work which has resulted in the load gaining potential energy. This, and other lifting machines, allow a small force (effort) to overcome or balance out a larger force (the load).

Experiment 13.16
The block and tackle

Another machine which is often used for lifting heavy loads is the pulley block, or the block and tackle. You may have seen one in use at your local garage when they are lifting the engine out of a car. You may have a block and tackle at school by which you could lift a pupil. Find the weight of a pupil who is willing to be lifted by the block and tackle. Attach a

Fig. 13.24

girder

block and tackle

strong sling

suitable newton balance to the other end of the rope or chain so that you can measure the force you have to exert to lift him. Measure the distance he is lifted off the floor and the distance the balance is pulled downwards. You can then calculate the work done on the pupil who is being lifted, and the work you have had to do to lift him. Which is greater, the work done by the pupil pulling (this is also the amount of energy he uses) or the potential energy gained by the pupil who was lifted?

In lifting the pupil by the machine you are doing something that you would probably not have been able to do unaided. The force you have to apply – the effort – is much less than the weight of the pupil – the load. However, your calculation will have shown you that the energy you have expended in lifting him is greater than the energy stored up in him as a result of the lifting. **Machines do not make energy**. In fact, there is always a 'loss' of energy when you compare the energy you get out of the machine with that you have put in. The energy is not, however, really lost. It is converted into some other form. When you lifted your partner off the floor with the block and tackle, where did this extra energy go?

Machines, therefore, simply make it more convenient for us to use our force to lift heavy loads which otherwise we might not be able to lift. They do not give us energy.

13.13 HOW DO WE EXERT FORCE?

In the last experiment you lifted a pupil off the floor by means of a block and tackle. What did you feel as you were doing this? Your muscles became taut, and you had to grip on the chain or the rope to exert a force. How is it that you are able to do that? Could you do it if you had no bones? Suppose you were just a mass of jelly, like a jelly-fish, do you think you would then be able to pull as efficiently as you can now? Of course, a jelly-fish might be able to crush an object by winding itself around it, but it is difficult to imagine such a creature being able to push a car along, or lift a car with a jack. It is interesting to see how plants and animals are able to support their own weight.

13.14 SUPPORT IN PLANTS

Experiment 13.17

Take two pieces of water weed and a spray of privit. Attach a lump of Plasticine to the base of each plant. Drop one piece of water weed into a tall beaker of water, and leave the other plants on the bench. Explain the differences in appearance between both pieces of weed.

Fig. 13.25 (a) and (b) show a section through the stem of a water plant and one through the stem of a woody plant respectively. With the help of these photographs explain why privet remained upright but the water weed collapsed. Examine the outside of the privet stem. What is around it? What is the function of this material? Peel off the outside layer. Hold the privet in your hand. Does it remain upright? The stem must have supporting cells inside it. Look back to Fig. 13.25. Where are the supporting cells arranged?

By looking at privet we can see that it is supported by woody material. Do all land plants have this form of support? The following experiment will help us to answer this question.

Experiment 13.18

You will require a flower pot containing four seedlings. Remove one seedling, A, and feel it. Is it hard? Rub the stem between your fingers. Is the stem easily broken or squashed? Do you think that there are any woody cells in the stem?

Remove a second seedling B and place it in a test-tube of water. Remove a third seedling C and leave it on your desk. Leave the fourth seedling D in the soil. Examine seedlings B, C, and D after about half an hour. What has happened to C? Have B and D changed in any way? Put seedling C into a test-tube of water. Remove the seedling after thirty minutes. Can you now say what is supporting this seedling?

Water enters a plant through the roots and evaporates from the leaves. Using this fact, explain why seedlings B and D remained erect but seedling C wilted when it was removed from the soil.

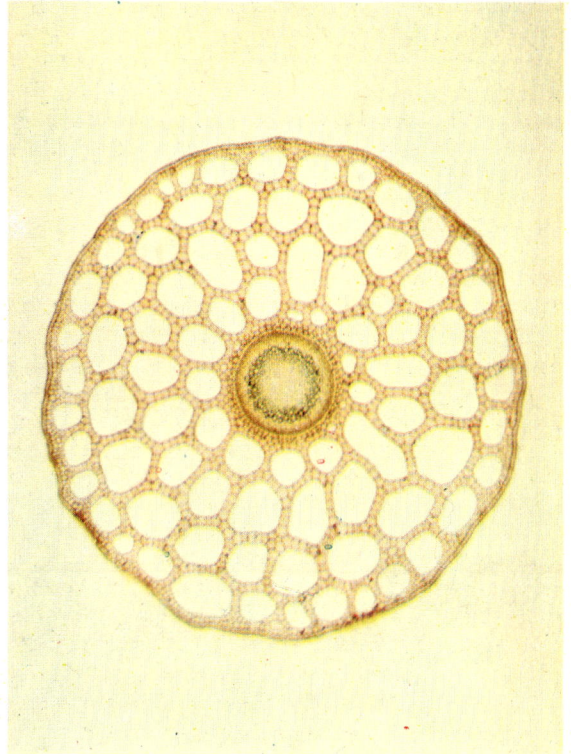

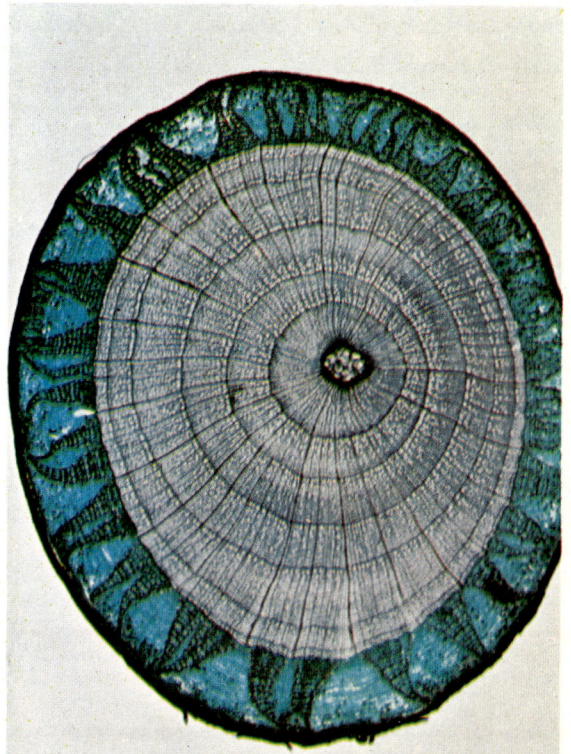

Fig. 13.25 Sections through (*above*) an aquatic stem, *Hippuris* (Mare's tail) and (*below*) a woody plant, *Tilia* (lime)

Experiment 13.19

Seedlings contain water. Plan an experiment to find the weight of water in a number of seedlings. Do all seedlings have the same percentage of water in them?

13.15 SUPPORT IN ANIMALS

In Unit 2 you divided animals into two groups. All the animals in one group had no internal skeleton, all those in the other group had an internal skeleton. What did you call these groups? Which group contained the larger number of animals?

13.16 SUPPORT IN INVERTEBRATES

The majority of animals have no internal skeleton. Can you suggest one possible means of support? Your observations in Experiment 13.18 should help you to answer this. Seedlings are supported by water in their cells. Can a liquid give support to invertebrate animals? We can find out if we use a 'sausage' balloon to represent the skin of an invertebrate such as the earthworm.

Experiment 13.20
A model worm

Fill a sausage balloon with water. Tie the open end of the balloon. Does the liquid support the skin?

Lay the balloon on your desk. With a piece of chalk, mark the position of the head and tail end of the balloon. Wrap your hand around the tail and squeeze gently. What happens? Give another use for the liquid inside the worm. Many invertebrates have an additional form of support.

Experiment 13.21

Blow up a balloon or fill it with water. Tie the end securely. Cover the balloon with papier mache. When the papier mache has dried, burst the balloon. What happens to the papier mâché? Does it collapse or retain its shape?

The papier mâché covering represents a form of support seen in many invertebrates, especially arthropods. This group of animals includes members of the insects, the spiders, the centipedes and the crab family. These animals have a supporting skeleton *outside* the body.

Experiment 13.22
External skeletons

Examine the outside of the following animals: grasshopper, spider, crab, fly, centipede, shrimp, and moth. Do all the skeletons feel hard? Choose a scale of hardness, e.g. represent a very hard skeleton by four ticks √√√√, and a skeleton that is not very hard by one tick √. Write down the names of the animals you examined and alongside each the degree of hardness of their skeleton.

13.17 SUPPORT IN VERTEBRATES

Vertebrates are supported by an internal skeleton. Examine Fig. 13.26 or a model of a human skeleton. Find the spine. This is the central support of the body. Is it straight or curved? Does it contain one long bone or many small bones? Run your finger down your spine. What are the small lumps you can feel? The skeleton is made up of a

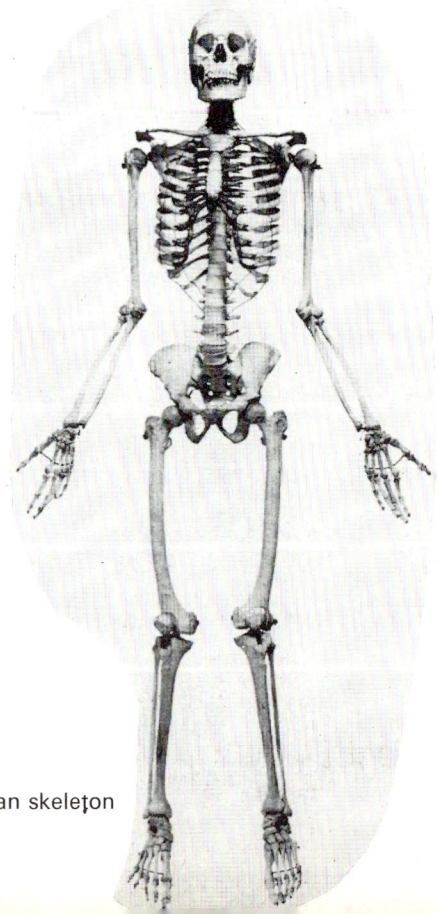

Fig. 13.26 A human skeleton

Fig. 13.27 Do these animals have hard external skeletons?

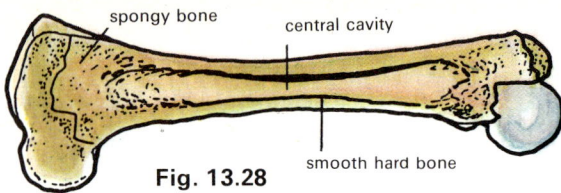

Fig. 13.28

large number of bones. Some are small, like those in the spine. Others are long like those in the limbs.

Fig. 13.28 shows a section through a long bone. Your teacher will show you a bone that has been cut through. Look for the smooth hard bone, the spongy bone, and the central cavity.

Experiment 13.23
Looking at bones

Bring the bone from the leg of a chicken to school. Compare it with the leg bone examined above. Write down two differences between the bones.

Look again at the human skeleton. Find where the arms and legs are attached. The arms are attached to large flat bones which form the shoulder girdle. The legs are attached to another group of large flat bones which form the hip girdle. The weight of the body is supported by these girdles. Figs. 13.29 and 13.30 show the

skeleton of a horse and that of a whale. Although the whale is a very large animal its skeleton is reduced so that it is light and flexible. In the horse the skeleton forms a rigid framework which supports the body. Can you explain why the whale does not need a large supporting skeleton? Your observations in Experiment 13.17 should help you to answer this. Find the shoulder and hip girdles on both these skeletons. Compare the relative size of these girdles with those on the human skeleton, and then answer the following questions:

1. In the horse is there much difference in size between the shoulder girdle and the hip girdle? In man is there much difference in size between the girdles? How is the size of the girdles related to the method of locomotion of these animals?
2. Why are whales able to grow larger than land animals?
3. Explain the following statement – 'If a whale is removed from water it will be crushed to death by its own weight'.
4. Why does the hippopotamus rarely leave water?

Experiment 13.24

Using apparatus in the laboratory, plan an experiment to show why animals that live in water can grow to a larger size and be heavier than land animals.

13.18 STABILITY

Land vertebrates are supported by their skeleton. Does the skeleton keep the body upright at all times? If you have ever watched someone try to skate for the first time you will know the answer to that question. There must be some other factor which is important in keeping an animal's body upright.

Fig. 13.29 Skeleton of horse (*left*)

Fig. 13.30 Skeleton of whale (*below*)

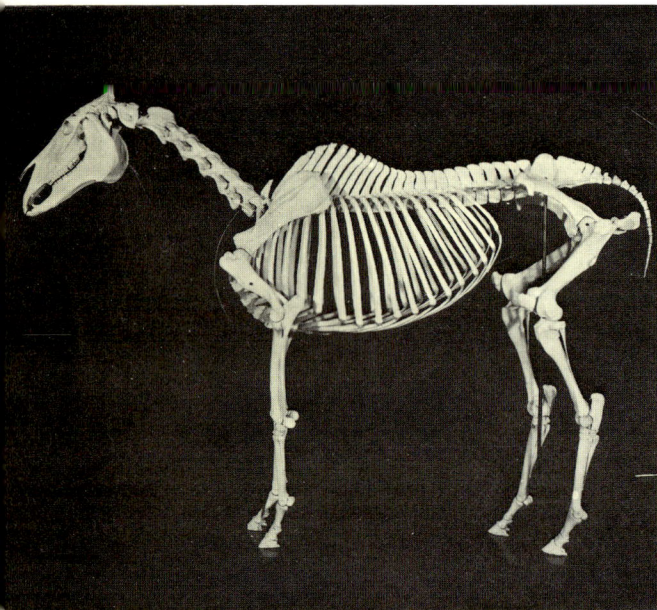

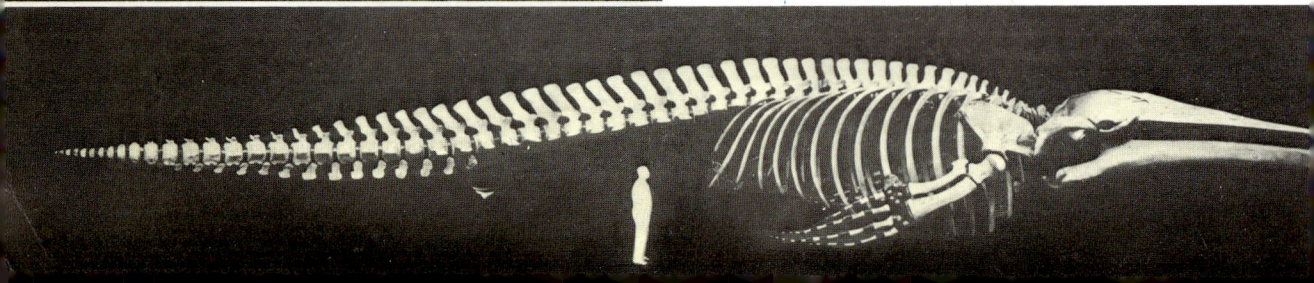

Experiment 13.25

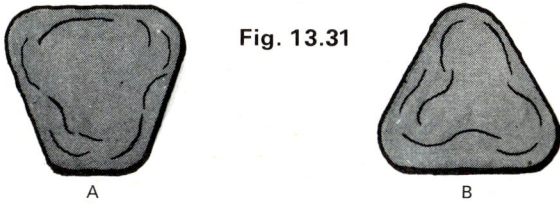

Fig. 13.31

A   B

Take two balls of Plasticine, each 20 g, and make the shapes shown in Fig. 13.31. Tap the top of each model with your finger. What happens to each? Model B is said to be **stable** because it remains upright. Model A is unstable. The more stable an object is, the more pushing and pulling it can withstand without falling over.

The limbs of land animals are used for support and movement, but they are also important in giving the animal stability. How do the limbs affect stability? This is the problem you are now going to investigate.

Experiment 13.26

Roll 50 g of Plasticine into a cylinder 4 cm long. This represents the body of an animal. Cut four straws, each 8 cm long. Insert one straw vertically below each corner of the model. Tie a string around the body and pass the string over a pulley attached to the end of the bench. Compare your model with Fig. 13.32. Measure *a* and *b* and calculate the area of the base of the model.

Place a strip of wood in front of your model to prevent it being pulled along the desk. Add weights to the pan until your model topples.

Find the number of grammes required to topple the model. What force is this? Change the position of the straws so that *a* and *b* increase in length (as in Fig. 13.32). Measure *a* and *b*. Add weights until this model topples. Enter your results in a table, like the one below.

Length *a* (cm)	Length *b* (cm)	Number of grammes required to topple model	Force required to topple model (N)

Which model is most stable? What is the relationship between the area of the base and the stability of the model?

Experiment 13.27

Using the same apparatus as in Experiment 13.26, attach straws of 16 cm and 4 cm to the most stable model. Find the number of grammes and calculate the force required to topple each model. Enter your results in a table.

Some questions for you to answer

1. Which length of leg gives the greatest stability?
2. Which is the most stable structure in terms both of arrangement and length of leg?
3. Why are the legs of the vaulting horse in your gymnasium splayed out?
4. Why do you prevent yourself from toppling forward if you stand with one foot several inches in front of the other?

Fig. 13.32

(b)

(a)

a = space between front legs
b = distance from front to back legs

13.19 BODY SHAPE AND STABILITY

Gravity pulls down on every molecule in an organism but it is convenient to think of it acting at one point in the body with a force equal to the weight of all the molecules. This point is called the **centre of gravity** or the **centre of balance**. If an object is suspended and allowed to swing freely it will come to rest in such a way that the centre of balance lies somewhere along the line that can be drawn vertically from the point of suspension through the object.

Experiment 13.28

Draw the shape of an animal, e.g. a fish, on a piece of cardboard. Cut out the shape. Make a small hole somewhere on the edge of the model. Suspend the model from a pin, and attach the pin to a board. Make sure that the model can swing freely. Allow the shape to come to rest. Attach a weighted string (a plumb line) to the pin so that it hangs in front of the model. When this comes to rest it will be vertical. Using the string as a guide, draw the vertical line on your model. The centre of balance of the model lies along this line. Can you think of a way of drawing another line on your model on which the centre of balance will lie? The centre of balance will be the point where these lines cross. Find the centre of balance of your model. How can you check that this point is the centre of balance?

Draw several body shapes such as a mammal with legs, a mammal without legs, a reptile such as a crocodile, and find the centre of balance of each.

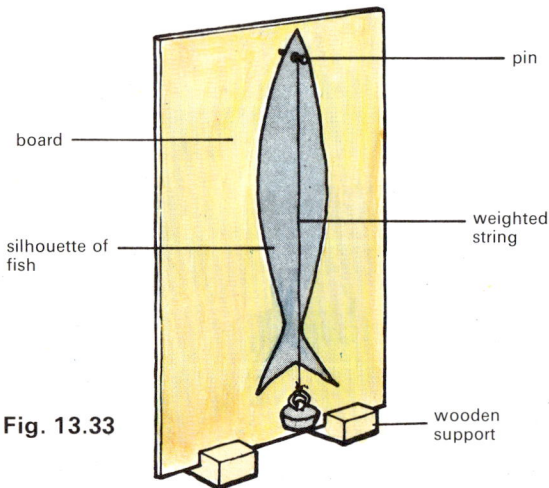

Fig. 13.33

Some questions for you to answer

1. Which body shape combined with which set of legs would give the most stable structure of all (i.e. the structure which does not easily fall over)?
2. How can an animal such as a giraffe change its stability?
3. If the centre of balance of an animal moves so that it is outside the area of its base, what will happen to the animal? Explain your answer.

Fig. 13.34 A giraffe walking along and eating. Is the animal's centre of balance the same in both pictures?

13.20 MUSCLES

Look again at the human skeleton. Examine an arm and find out where the arm can bend. You should see at least three regions where bending can take place. These regions are called **joints**. There is a joint at the shoulder, and at the elbow, one at the wrist and several between the bones of the fingers. Bones therefore move at a joint. But what causes the bones to move? They cannot move by themselves. The tissue in your limbs is mainly **muscle** tissue, and it is muscles that move the bones in the body. Fig. 13.35 shows two muscles that are found in the upper arm. How are these muscles attached to the bones?

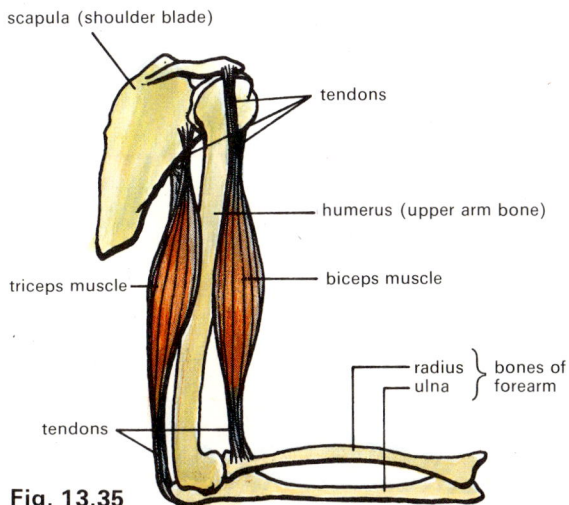

scapula (shoulder blade)

tendons

humerus (upper arm bone)

triceps muscle

biceps muscle

radius ⎱ bones of
ulna ⎰ forearm

tendons

Fig. 13.35

Experiment 13.29

Stretch out your arm and rotate it so that the palm is uppermost. Grip the upper part of the outstretched arm with your other hand. Slowly bend the outstretched arm from the elbow. What change do you feel in your upper arm? What muscle are you feeling?

The biceps muscle shortens and in doing so lifts the forearm. A muscle works by contracting. Once it has contracted it will remain in that position. It cannot elongate and push. How then do we lower our arm? Look again at Fig. 13.35 and find the triceps muscle. What will happen to the forearm when the triceps contracts? Will the biceps change in length after this contraction? If

it does, how will it change? The muscles that move our limbs occur in pairs. Contraction of the biceps lifts our forearm, contraction of the triceps lowers the forearm. In what other parts of the body would you expect to find pairs of muscles that work in this way?

Experiment 13.30
How much can your biceps muscle lift?

Tie a length of string around each of several bricks. Form a loop, large enough to go over your hand, in the end of the string away from the brick so that the distance between the loop and the brick is approximately the height of your desk. Sit at your desk so that your elbow is resting on the edge of the desk, but your forearm has no support. See how many bricks you can pick up,

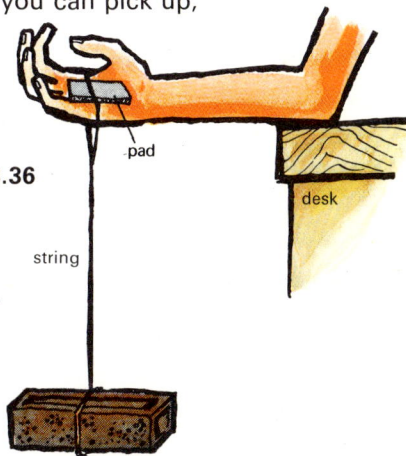

Fig. 13.36

pad

desk

string

(a) by placing the string over your hand,
(b) by placing the string 5 cm from your hand,
(c) by placing the string 15 cm from your hand.
Weigh a brick then calculate the total weight lifted by each part of the arm. Which part of the forearm can lift the greatest weight? What does the arm act as? In this experiment what is the load? What is the effort? What is the fulcrum?

Experiment 13.31
What force is exerted by the biceps in lifting a load?

We can find out the force exerted by the biceps in lifting a load if we use a spring balance to represent the muscle. Using the apparatus

Fig. 13.37

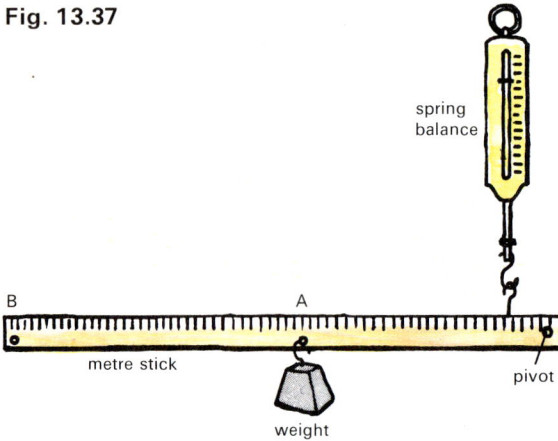

SUMMARY OF UNITS

Quantity measured	Unit used	Size	Everyday use
Force	1 newton (N)	10 N ... 1 kg	weight of an average dessert apple ... 1 N
Energy	1 joule (J)	$1 N \times 1 m$ ... 1 m	1 N m = 1 J ... energy to lift apple from floor to table
Power	1 watt (W)	1 J in 1 s ... 1 m	1 s ... power required to lift apple to table in 1 second

illustrated in Fig. 13.37 attach a weight to point A. What is the reading on the spring balance? Move the weight to point B. What is the reading on the spring balance?

Repeat this experiment using different weights. Make a table of your results. What conclusions can you draw from this experiment?

WHAT YOU HAVE LEARNT IN THIS UNIT

1. Pushing, pulling, twisting, turning, deforming, stretching, and compressing are examples of kinds of forces. Friction and gravity are two forces which we come across in nature.

2. When there is no friction or effective opposing forces (i.e. if there is no unbalanced force acting), moving objects will continue in steady motion.

3. An extra, or unbalanced force acting on an object will cause it to accelerate.

4. Hovercraft ride on a cushion of air, which reduces the effect of friction.

5. Each time an equal extra weight is hung on a spring (up to a certain limit) an equal extension is produced. We can make use of this fact to measure forces. The spring balance depends on this principle.

6. Forces are measured in units called newtons. On the planet Earth, the weight of a 1 kg mass is approximately 10 newtons. On the Moon the weight of a 1 kg mass is about 1.6 newtons.

7. Levers can magnify the effect of a force. The force required to lift a certain load is called the effort. The effort has to be moved through a greater distance than that through which the load moves, so although a small force can lift a heavy load it has to move through a greater distance to make up for it.

8. A small force has more turning effect the further it is from the pivot.

9. Forces occur in pairs, which act in opposite directions to each other.

10. Land plants are kept upright by strong woody fibres in their stems, which give them support. Some plants are supported by water in their tissues.

11. Animals which have no internal skeleton (invertebrates) may be supported like plants by water, or they may have an external skeleton, like a shell.

12. Vertebrates have an internal skeleton which is made up of many bones jointed together for flexibility. The strength of the bones depends on the weight and needs of the animal. Birds have thin hollow bones to decrease their weight so that they can fly more easily.

13. The stability of an animal, i.e. whether it will fall over easily or not, depends on the position of its centre of gravity.

14. In lifting objects we use our muscles. The muscles that move our limbs are in pairs. A muscle works by contracting; it cannot elongate and push. Once contracted it can only be restored to its original condition by the action of the paired muscle, which, in doing this, contracts.

Unit 14
Transport Systems in Living Things

14.1 WHAT ARE WE TALKING ABOUT?

What are transport systems? As soon as you hear or read the word 'transport' you probably think of buses and trains and aeroplanes. Yes, these are transport systems. The word 'transport' means 'carrying from one place to another'. Although these methods of getting about from one place to another are very important in our everyday lives, they are nothing like so important as the transport systems in our bodies. What is carried from one place to another in our bodies? Food for one thing; oxygen from the air for another; and waste materials which are left over from our food or are produced as the result of our using food for obtaining energy.

In this unit we want to find out how these materials are transported from one part of our body to another. By the end of it you will be impressed by the wonderful way in which we are made and in which we function.

14.2 TYPES OF FOOD

Let us begin by thinking about food. There are four main kinds of substances which make up our food – starch, sugar, protein, and fats. Perhaps you do not recognize any of your familiar foods in this list – things like corn-flakes, bread, butter, meat, and ice-cream – but everything you eat can be classified under one or more of these headings. You have already worked with starch and the sugar called glucose, and you have found in Unit 5 (Book 1, page 90) that starch could be changed into a sugar similar to glucose. Starch and glucose, you will remember, belong to the same class of substance, which we call carbohydrates.

Proteins, fats, and carbohydrates all play an important part in keeping you fit and healthy. You will discover later in this unit what each does, but first we must find a way to test for these substances in the foods we eat.

Fig. 14.1 A transport system. Can you relate this to your own body and the movement of liquids around it?

	Starch	Glucose	Protein	Fat
Place a drop of each liquid on a filter paper. Allow each spot to dry. Does an oily mark form?				
Add drops of iodine solution to each sample in turn.				
Add 5 drops of Millon's reagent (care) to a sample of each foodstuff. Heat carefully in a beaker of boiling water for about 3 minutes.				
Add 1 dropperful of Benedict's solution to a fresh sample of each. Heat carefully in a beaker of boiling water for about 3 minutes				

Experiment 14.1
Testing for foodstuffs

Take about 1 cm³ of each of the four substances provided, and place each in a labelled test-tube. Carry out the tests listed above and note what happens.

Look at the results you have recorded in your table. What do you think are the best tests for starch, glucose, protein, and fat?

Now that you have found a quick and easy way to test for starch, glucose, protein, and fat, you can find out which of these substances are present in a variety of foods that you eat each day.

Experiment 14.2
What substances are present in our food?

Grind up some potato in a little water. Pour some of the potato juice into each of four labelled test-tubes and find out what food substances are present by applying the tests you have discovered in Experiment 14.1. Enter your results in a table.

Test other foods in a similar way. Try to include samples of the foods you had for breakfast or lunch today.

For proper growth and for healthy living it is necessary for us to include in our diet sufficient of each type of compound – carbohydrate, protein, and fat, and as you will see later on, other things as well. A diet which includes the right proportions of each of these is called a balanced diet. If your food contains a lot of carbohydrate and very little protein it would not be balanced.

Some questions for you to answer

1. Name two foods which are rich in starch.
2. Which food that you tested was rich in fat?
3. Name one food which contains protein but no starch.
4. Does cane or beet sugar contain glucose?
5. What type of compound – carbohydrate, protein, or fat – was most common in the food you had for breakfast or lunch today?
6. What foods would you include in a balanced diet?
7. Do you consider that today's lunch was a well-balanced meal?

Items in our diet	Starch	Glucose	Protein	Fat
Potato				

14.3 ENERGY RELEASE

Energy from the sun is stored in food. What is the name given to this process? If you cannot remember, turn back to Unit 8, Book 1. This energy is released in organisms during respiration. Do all foods provide us with the same amount of energy, or do some give us more than others? This is the problem that we now have to investigate.

To do this we must be able to measure the energy released. Energy occurs in many different

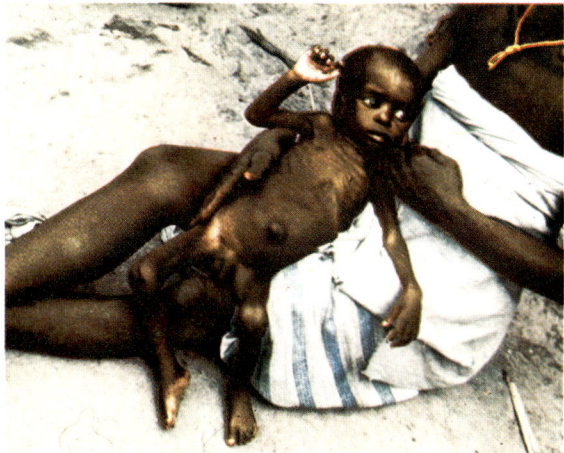

Fig. 14.2 A child suffering from protein deficiency

forms, as you know. Which of these is the easiest to measure? If you thought it was heat energy, you were correct. You know that we use a thermometer to detect heat energy, but can we compare the heat released from an apple with that released from a slice of bread by holding a thermometer above each? Obviously we cannot. Can you think of a way to measure the heat released from a piece of bread when it is burnt? You should be able to begin planning this experiment by answering the following questions.

1. How can you release the heat energy quickly from the bread?
2. Can this heat energy be used to heat something else – if so, what?
3. How can you reduce the amount of heat energy lost to the atmosphere?
4. With what gas will you surround the food to make sure that all the heat energy is released?

Look at the following instructions, and see if you have answered the questions correctly.

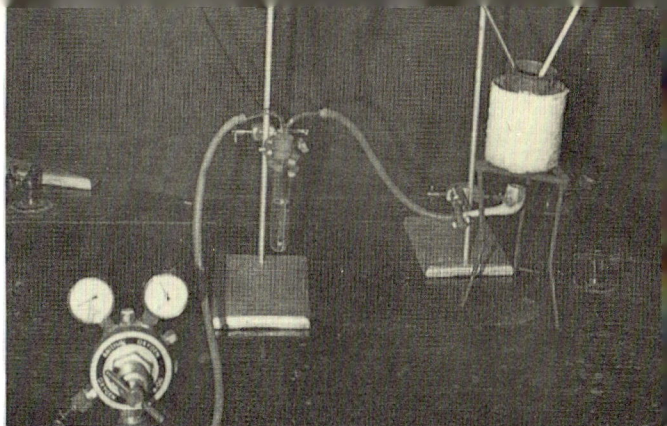

Fig. 14.3 A food calorimeter

Experiment 14.3

Using the apparatus shown in Fig. 14.3, put 100 cm³ of water in the metal beaker and record the temperature of the water. Put 1 g of dried bread in the bowl of the pipe, and your teacher will adjust the flow of oxygen through the pipe. Ignite the bread. Stir the water in the beaker, but take care not to hit the thermometer as it is easily broken. Record the highest temperature of the water.

By how much has the temperature of the water increased? Heat energy released from the burning bread has increased the temperature of the water. Was all the heat energy from the bread used for this purpose? Obviously not, because some of the heat was used to warm up the air round about the burning bread. Nevertheless, the experiment will give us some idea of whether we can get a lot or a little heat energy from foods.

Repeat the experiment using different foods. Will it be necessary to weigh the food you use? If so, what weight will you use? Why did you decide to use this weight? Enter your results in a table showing what rise of temperature you get when using different foods.

Which foods gave out most heat energy? Look back at the table on page 86. What was the main type of compound in these foods? Why were dried foods used in this experiment? Perhaps Experiment 13.19 (page 78) will help you to answer this question.

Foods rich in fats and carbohydrates provide us with the energy we need. Only part of these foods is broken down to give us energy. The rest is stored under the skin and on certain organs in the body as fatty material. If we eat too much of foods rich in fats and carbohydrates what happens to us?

Look again at the results of Experiment 14.3.

Do foods rich in protein liberate much energy? They are important to us in other ways. They provide us with material to make new cells. What might these new cells be used for?

14.4 OTHER SUBSTANCES IN FOOD

In addition to carbohydrates, proteins, fats, and water, foods contain small amounts of vitamins and mineral salts. Both of these substances must be present if the cells in our bodies are to work properly. Mineral salts are also used in the formation of certain tissues in the body, for example bone and blood cells.

Certain foods, especially vegetables, contain a substance called cellulose. This substance cannot be digested by man, but it is important in providing bulk to our food. You will see later the importance of this bulk.

Something for the girls to do

Plan a balanced menu for breakfast, lunch, and dinner for

(a) a girl who has to eat lunch at her office;
(b) a girl who wants to lose weight over a period of weeks.

Something for the boys to do

Plan a balanced menu for breakfast, lunch, and dinner for

(a) a thin boy who wants to play rugby and must develop muscles but not fat;
(b) a fat boy who wants to be picked for the athletic team.

14.5 FEEDING IN ANIMALS

Feel the teeth in your mouth with your tongue. Are they all the same shape? How many different shapes can you feel? How many teeth have thin, sharp edges? How many are pointed and how many feel like double teeth? Do all the teeth in your mouth carry out the same function? This question can be answered by the following experiment.

Experiment 14.4

Eat an apple. Which teeth did you use to (a) bite the apple, (b) chew the apple? From a bar of toffee, bite off a small lump and chew it. Did you use the same teeth to bite the toffee as you did to bite the apple?

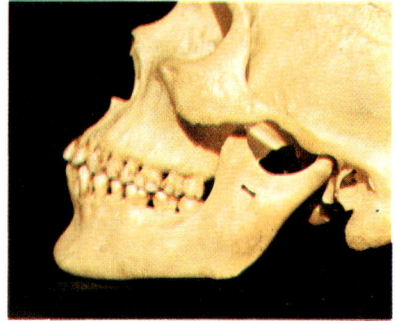

Fig. 14.4 A human skull

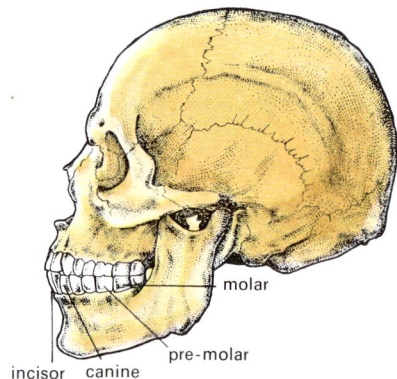

Fig. 14.5

You probably used your front teeth or **incisors** to bite the apple, and the more pointed teeth or **canines** to tear the lump from the bar of toffee. You chewed both with the large double teeth or **molars** at the back of your mouth. Find these teeth in Fig. 14.5. You will see that the molars have been split into two groups; the back three on each side are called molars, the teeth between them and the canines are called **premolars**. How many molars are shown in the diagram? How many molars do *you* have?

Fig. 14.6

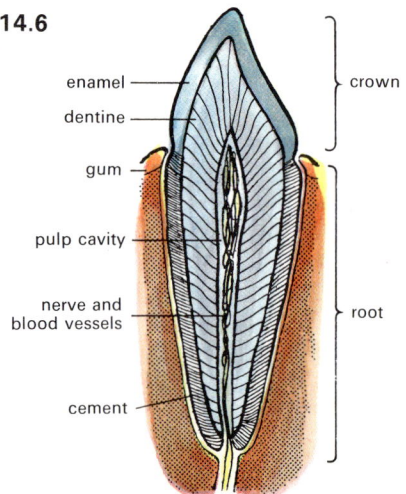

14.6 TOOTH STRUCTURE

Fig. 14.6 shows a section through a human incisor. Each tooth consists mainly of a bone-like material called **dentine**. The crown of the tooth is protected by a very hard substance called **enamel**, and the root is covered by a layer of **cement**. The pulp cavity in the middle of the tooth contains nerves and blood vessels.

14.7 CARE OF TEETH

Our teeth, although very important to us, are often neglected and abused. Are all your teeth in perfect condition or have some had to be filled or extracted? Make a survey of your class and for each pupil find the number of teeth that have fillings in them and the number that have been extracted. How many pupils in your class have got perfect teeth? Compare your survey with those carried out by other classes. How can the results be presented to enable you to do this quickly?

How many pupils in your year have got perfect teeth? It has been estimated that in this country two people in every hundred have perfect teeth. How do your results compare with this estimate?

14.8 WHAT CAUSES TOOTH DECAY?

Experiment 14.5

Take a tooth and cover it with wax. Allow the wax to harden, then scrape away some of the wax. Place the tooth in dilute hydrochloric acid for three days. Remove the tooth and scrape off the wax. Prod the tooth with a needle. What has the acid done to the enamel of the tooth?

We now know that acid removes enamel from teeth. Where does the acid in our mouth come from? The following experiment will show you.

Experiment 14.6

Squirt the saliva in your mouth several times through the spaces between your teeth. Collect this saliva in a test-tube. Label three test-tubes A, B, and C. Into A put 1 cm³ of saliva, into B put 1 cm³ of glucose solution, and into C put 1 cm³ of both saliva and glucose solution.

Leave the tubes in an incubator set at 37 °C for one or two days. Then remove samples from each tube and test each with pH paper or litmus paper. What has happened in tube C? Why did you set up tubes A and B? What do we call these tubes?

Sweet, sticky, particles of food become trapped between our teeth and these are changed to an acid by bacteria in our mouth. How could you find out if bacteria are present in your mouth?

When a small hole has been formed in a tooth more food collects in it and this in turn is changed to an acid. The hole gets larger and larger. The dentist can save the tooth by removing the decayed part. He prevents further decay by adding a protective filling to the cavity. However, if the tooth is not treated soon enough it has to be extracted.

14.9 PREVENTION OF TOOTH DECAY

We can do two things to prevent tooth decay, remove the cause of the decay and strengthen the teeth. The formation of acid can be reduced by brushing the teeth. This process dislodges particles of food and helps to remove bacteria from the teeth. Here is an experiment to try.

Experiment 14.7

Put 1 cm³ of toothpaste into a test-tube containing 10 cm³ water. Shake the tube vigorously for a few minutes. Add pH paper. Is the liquid acid or alkaline? Remove a drop of liquid from the tube and observe it under the low power of the microscope. What do you see? From your observations write down *two* ways in which toothpaste may help to prevent tooth decay. Which of these do you think is more important? It may help you to answer this question if you remember that toothpaste is in contact with the teeth for only a short time before being rinsed away.

Experiment 14.8

Test a variety of toothpastes and toothpowders as in Experiment 14.7. From your results which do you think is best?

Something to do at home
Design a poster advertising your brand of toothpaste. In your design show why you selected that particular brand.

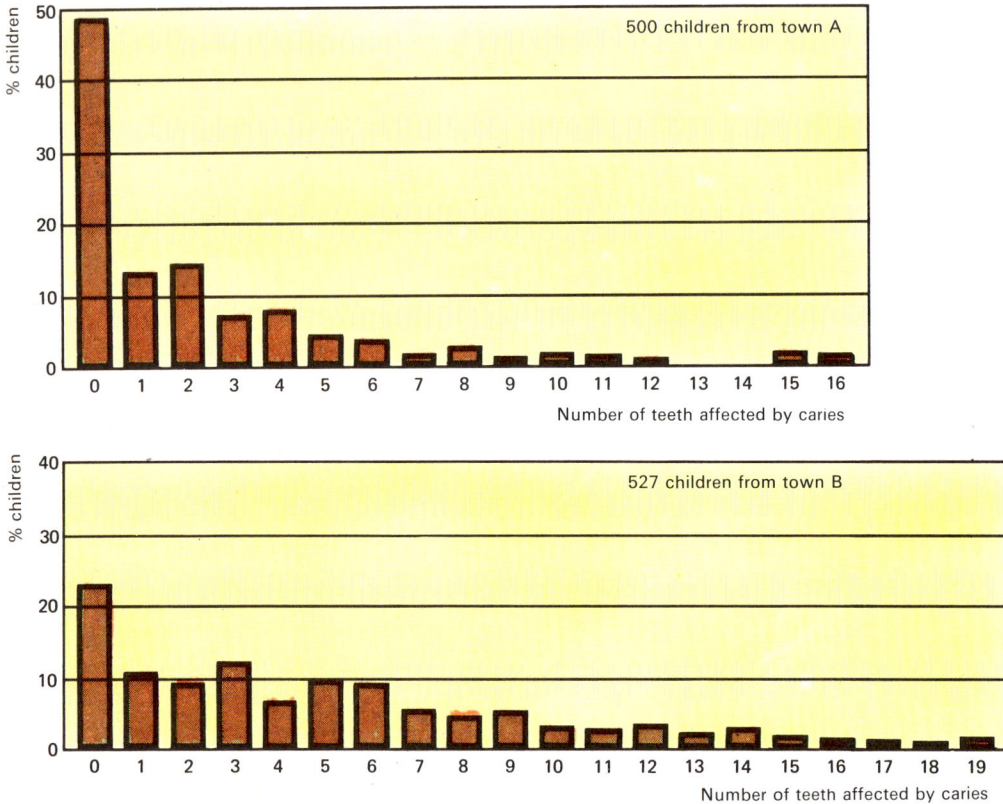

Fig. 14.7

14.10 STRENGTHENING THE TEETH

Surveys similar to the one you conducted in class to find the amount of dental decay have been carried out on a large scale in certain towns in Britain. The results obtained are shown in the diagrams above, Fig. 14.7.

Look at the histograms and answer the following questions:

1. What percentage of children in town A had perfect teeth?
2. What percentage of children in town B had perfect teeth?
3. In both towns a few children had a large number of teeth affected by decay. What was the greatest number of teeth affected per child in each town?
4. In which town had the children the strongest teeth?

How can we explain these results? You could say that the children in A brushed their teeth regularly, while those in B did not. But is this a likely explanation? Obviously this is not the

answer. The children in A drink water from the same reservoir, and those in B drink water from another source. Could something in the drinking water strengthen the teeth? You could argue that drinking water is rain water and will be the same in both reservoirs. However, when water flows over the ground certain chemicals are dissolved in it. One such chemical is a fluoride. When the water from towns A and B was compared it was found that a trace amount of fluoride was present in the water in town A but not in town B. We cannot say from one survey that fluoride reduces the amount of tooth decay. However, many surveys have been carried out and all show similar results.

Sodium fluoride can be added to drinking water at the waterworks, but many people have fought to prevent this. They argue that their drinking water should not be interfered with. But is the water you get from the tap pure rain water in any case? What happens to it before it reaches your home?

Some homework

1. You have a friend who thinks that fluoridation is wrong because it contaminates drinking water. Write a letter to reassure him that this process is not harmful and point out the advantages of fluoridation.

or

2. Compose a letter which could be sent to a firm which manufactures products for tooth care and suggest ways in which they could use fluoride.

or

3. Try to find out the names of towns which have fluoride added to their drinking water.

14.11 TEETH IN OTHER MAMMALS

Our hands play an important role in feeding. If you hold your hands behind your back as you try to eat a meal you will realize how important they are. Have you ever wondered how animals which have no such specialized limbs eat? This is the problem you are now going to investigate.

Fig. 14.8 The teeth of two carnivores — a lion and a dog

Experiment 14.9

Look at the skulls of a dog and a rabbit. With the help of Fig. 14.9 identify the teeth in each

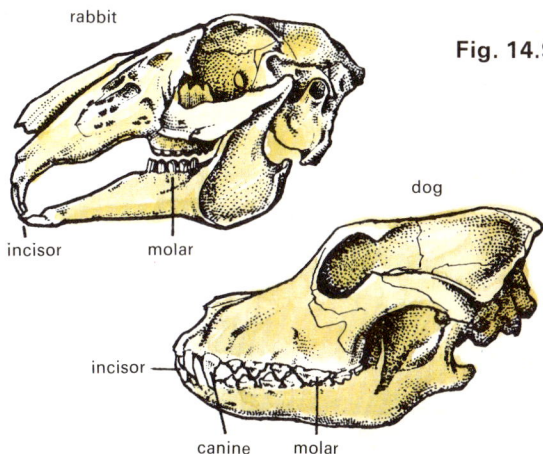

Fig. 14.9

skull. Run your fingers over the teeth at the back of a sheep's skull. Are these teeth smooth or ridged? Look at the shape of the teeth in the skull and decide how they are used in feeding. When a sheep is chewing grass do the jaws go up and down or from side to side? Hold the lower jaw against the upper jaw and see how the teeth in each jaw fit together. Move the lower jaw from side to side. When a sheep chews grass it makes this action with its jaws. Can you name a process in which plant material is ground by being placed between two stones which move from side to side over each other?

Look at the dog's skull. How do you think the canine tooth got its name? Suggest how each type of tooth is used in feeding. Hold the lower jaw against the upper jaw. Try to move the lower jaw from side to side. Can you do this easily? In what direction do the jaws move when a dog eats?

Fig. 14.10 shows the skull of an unknown animal which was found in a field. Look at the teeth and decide if this animal was a carnivore or a herbivore. Give reasons for your decision.

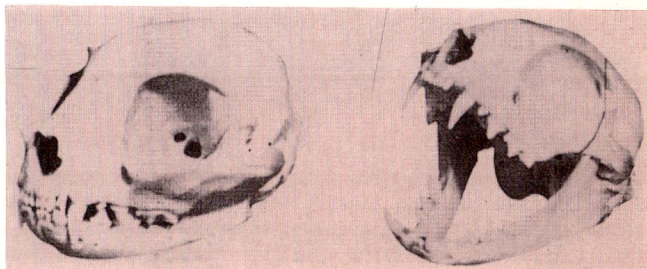

Fig. 14.10

14.12 FEEDING THE INVERTEBRATES

1. The locust

Do insects have teeth similar to ours? The locust is a herbivore, does it have teeth like a sheep? Does it have teeth at all?

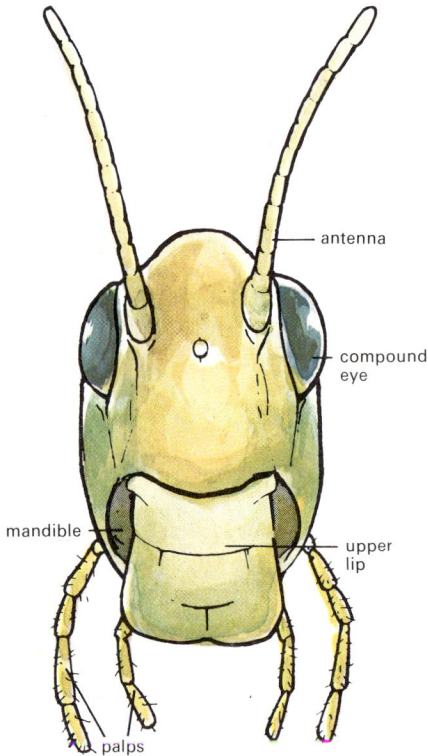

Fig. 14.11 Mouth-parts of a locust

Experiment 14.10

Your teacher may be able to give you a locust in a jar. Watch this locust feeding. If you have no locusts in your laboratory watch a film showing a locust feeding. With the help of Fig. 14.11 answer the following questions:

1. In which direction does the upper lip move?
2. In which direction do the jaws or mandibles move?
3. How does the locust use the palps?
4. How are the front legs used in feeding?

Your teacher will dissect the mouthparts of a dead locust. Feel the mandibles. What do you think they are used for? Why do you think the locust is called a biting insect?

Experiment 14.11

It has been estimated that a swarm of locusts contains about a thousand million insects. Find the weight of grass eaten by the locusts in your laboratory in a day. Using this information, work out the weight of grass that would be eaten by one thousand million locusts in a day.

In Unit 2 (Book 1) you examined another insect which eats by biting leaves. What was this insect called? Watch this insect feed. Does it have mandibles?

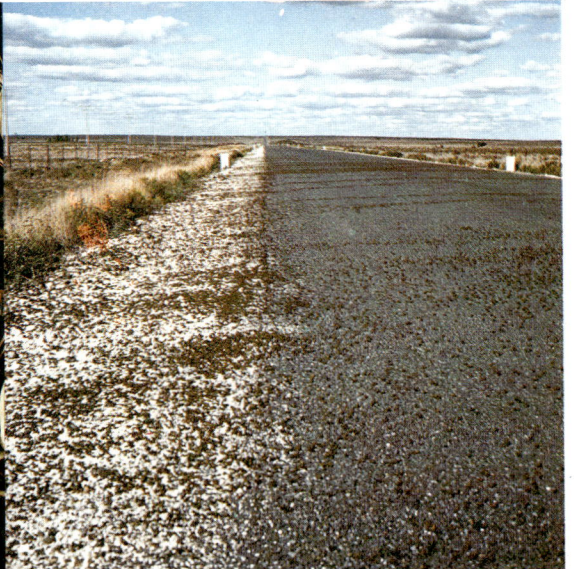

Fig. 14.12 (*left*) Locusts devouring a crop and (*above*) a swarm of locusts

2. The housefly

Experiment 14.12

Catch a housefly and leave it for a day in a jar covered with muslin. Then put a few drops of sugar solution into the jar and with the aid of a lens observe the fly feeding. Look at it from below by lifting up the jar. Observe the legs of the fly. Do they remain still while the fly eats?

At the front of the fly's head is a tube with a sucking pad at the end. The fly can eat food only if it is liquid. When it alights on solid food, e.g. sugar, saliva passes from the feeding tube on to the food. The food is dissolved by the saliva and then sucked up the tube. The saliva may contain many harmful bacteria and in this way flies spread disease. From your observations in Experiment 14.12 write down another way in which flies spread bacteria.

3. The mussel – yet another method of feeding

Experiment 14.13

Your teacher will set up a binocular microscope and a microscope. Under the binocular microscope you will see a mussel opened to display the large gill flaps, and under the microscope you will see a small piece of tissue cut from a gill. Small mud particles have been dropped on to the gill flaps. Look down both microscopes then answer the following questions:

1. Are the mud particles still or moving?
2. If they are moving, in which direction is this movement?
3. From your observations of the gill, what has brought about this movement?
4. Suggest what the mussel eats when it is in the sea, and how the food is brought to the gills.

Mussels live attached to rocks, and therefore do not search for food. Water containing small food particles is drawn into the animal. The food is picked up by the gills and carried to one end of the mussel. The food is then passed into the mouth which lies between the gill slits. Particles in the water that are not food are carried by the gills to another opening and out of the shell. The mussel is called a **filter feeder**.

14.13 DIGESTIVE SYSTEMS

In most animals digestion takes place in a tube called the **alimentary canal**.

Experiment 14.14

Put a small pond animal, e.g. *Daphnia*, or a small white worm on a slide. If you use a pond animal remove most of the water to prevent it moving around. Do not cover it with a coverslip. If you use a white worm use a coverslip. Look at the organism under the low power of the microscope. Fig. 14.13 will help you to find the intestine, which is part of the alimentary canal. What colour is the intestine? What do you think the animal eats? Give a reason for your answer. Look carefully for movements of the intestine wall. What do you think is the purpose of these movements?

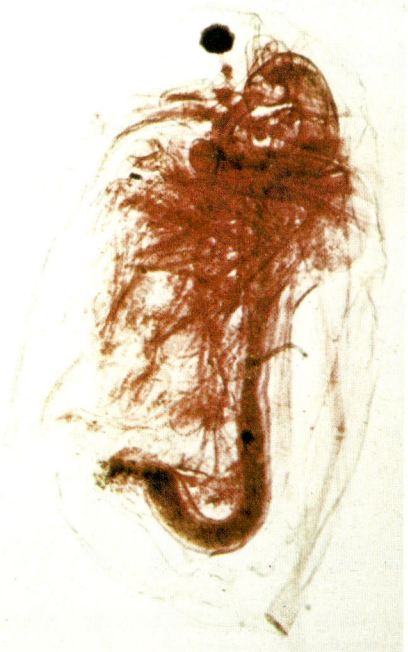

Fig. 14.13 *Daphnia*, a small pond animal

We shall now look at the alimentary canal of an animal you studied in Unit 2 (Book 1). The earthworm is much larger than the organisms studied in the experiment above. Does this mean that the alimentary canal will be more complex? Let us find out.

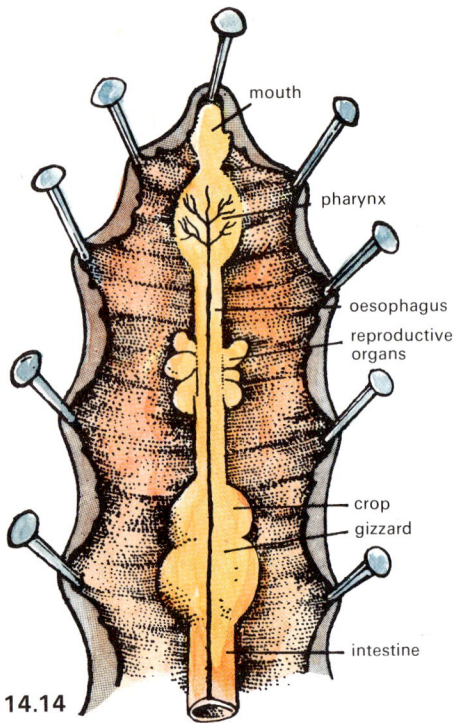

mouth
pharynx
oesophagus
reproductive organs
crop
gizzard
intestine

Fig. 14.14

Experiment 14.15

Your teacher will dissect an earthworm. Look at the dissection and from Fig. 14.14 find the following regions — mouth, pharynx, gullet (oesophagus), crop, gizzard, intestine, and anus. These are all regions of the alimentary canal. Although the canal looks more complex than that of *Daphnia*, it is still basically a long tube. Certain regions of the tube have become modified to perform a particular function. Feel the gizzard. What do you think is the function of this region? Give a reason for your answer. From your studies of the earthworm, what would you expect to find in the intestine? Open up the intestine and see if you are correct.

Fig. 14.15 A longitudinal section through part of an earthworm

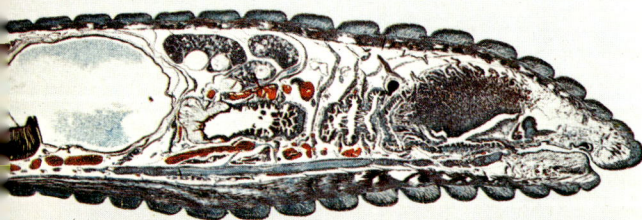

14.14 DIGESTIVE SYSTEM OF A SMALL MAMMAL

Experiment 14.16

With the aid of Fig. 14.16 find the following regions in the dissected rat: mouth, stomach, small intestine, large intestine, rectum. These are all regions of the alimentary canal. The tube joining the mouth to the stomach, the gullet or oesophagus, lies below the heart.

By means of blunt forceps unravel the intestine. What can you see in the clear tissue between the loops of the intestine? Remove the alimentary canal by cutting through the oesophagus and the rectum. Lay it on a dissecting board. You will see that again we have one long tube. Some regions are wider than others. Name a wide region. The large green structure, the caecum, contains bacteria which digest some of the food in the intestine. Although some digestion takes place in the stomach, most food is digested in the small intestine. Would you expect that part of the alimentary canal to be short or long? Why? Measure the entire canal, then the small intestine. Were you correct?

Fig. 14.16 Dissection of a rat, showing the intestine

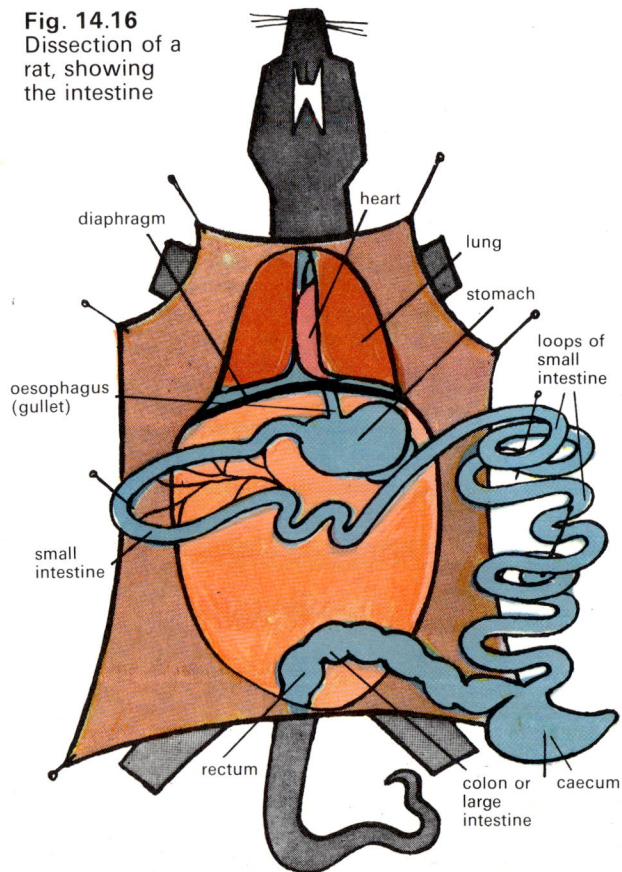

diaphragm
heart
lung
stomach
loops of small intestine
oesophagus (gullet)
small intestine
rectum
colon or large intestine
caecum

When you unravelled the intestine you cut through some tubes which joined it to the liver and the pancreas. These glands, and the salivary glands in the head, produce substances which pass into the alimentary canal and are used in the process of digestion. You have already examined one of these substances. What was it called? What role did it play in digestion? The alimentary canal and the organs and glands associated with it form the **digestive system**.

You will be provided with a diagram of the human digestive system. Name as many parts of this system as you can. Write down one difference between the alimentary canal of a rat and that of man.

By answering the following questions write a paragraph about digestion:

1. Where does food enter the alimentary canal?
2. What role do teeth play in the process of digestion?
3. Why must food be digested?
4. What substances play an important part in digestion?
5. Where does digestion take place?
6. How is food moved along the alimentary canal? (Your observations in Experiment 14.14 should help you to answer this.)
7. Why is roughage important?

14.15 TRANSPORT SYSTEMS

The small food molecules produced as a result of digestion, pass into the cells lining the intestine and then into the blood vessels in the intestine wall. Blood carries the food from the intestine around the body. Do you remember seeing blood vessels in the clear tissue between the loops of the small intestine? Blood transports food from one region of the body to another. It is part of a **transport system**.

14.16 TRANSPORT SYSTEMS IN PLANTS

Experiment 14.17

Stand some plant stems overnight in water dyed red. Remove a stem and cut 1 cm from the base of the stem. Using a hand lens look at the base of the cutting. Draw a diagram to show the places where you clearly see the dye.

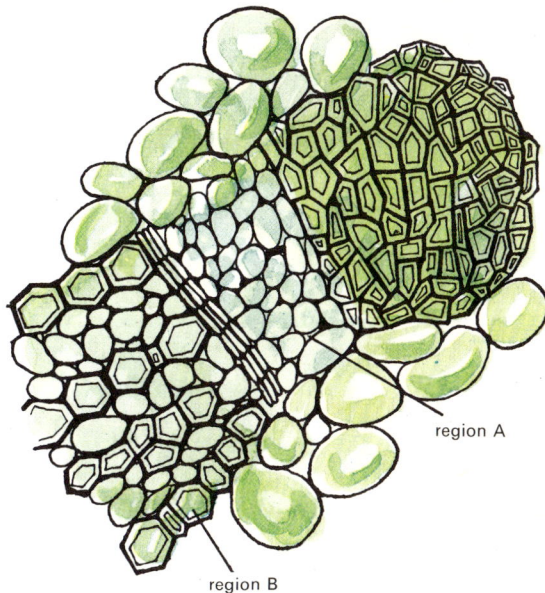

Fig. 14.17 Vascular bundle from a section across the stem of a non-woody plant

Fig. 14.17 shows a thin section of the stem of a plant. Your teacher will prepare a similar section from a stem that has been standing in coloured water. Look at this section under the microscope. Is region A or B stained red? Make a vertical cut down the stem. Draw a

Fig. 14.18 A vascular bundle from a section through the stem of *Curcurbita* (marrow)

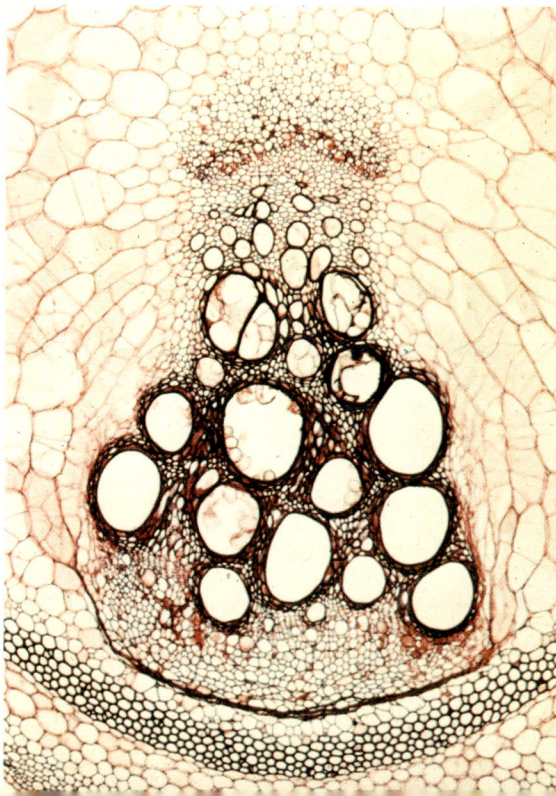

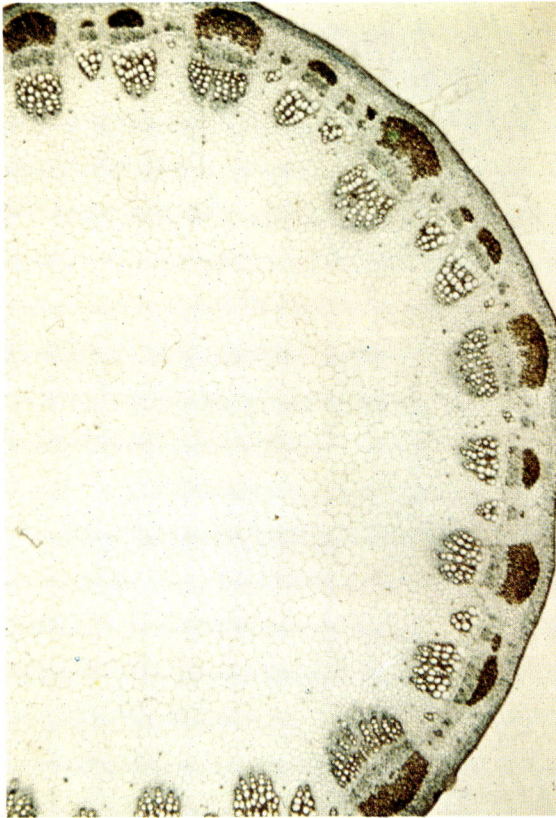

diagram to show the places where you clearly see the dye. With a pin and a pair of fine forceps carefully remove a coloured region. Fig. 14.20 will show you how to do this. Examine the coloured strands under the low power of the microscope. What do you see?

Now look at a whole plant that has been standing in coloured water for a day. Examine the roots, the stem, and the leaves of this plant and suggest the path taken by water as it moves through the plant.

Water enters the roots and travels up the stem to the leaves. The water is transported through a series of long thin tubes which run the length of the plant. Did you see any signs of strengthening material in these tubes when you examined them under the microscope? If you did, describe the way in which this material was laid down.

Fig. 14.19 Section through the stem of a non-woody plant, *Helianthus* (sunflower)

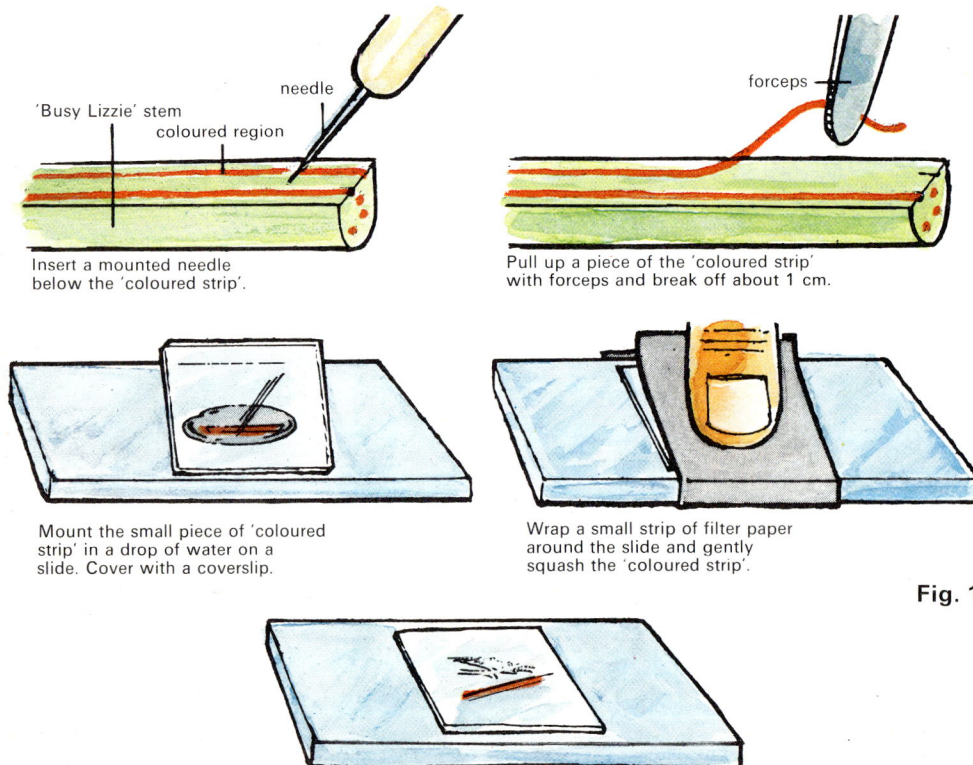

Insert a mounted needle below the 'coloured strip'.

Pull up a piece of the 'coloured strip' with forceps and break off about 1 cm.

Mount the small piece of 'coloured strip' in a drop of water on a slide. Cover with a coverslip.

Wrap a small strip of filter paper around the slide and gently squash the 'coloured strip'.

Fig. 14.20

Experiment 14.18

Using cuttings from a 'Busy Lizzie' plant find the rate at which coloured water moves up the stem. Test the following hypotheses:

1. Water will take longer to move up the stem if the leaves are removed from the stem.
2. Water will move up the stem faster if warm air is blown on to the leaves.

14.17 TRANSPORT ON A MINI SCALE

Experiment 14.19

Your teacher has set up two microscopes A and B. Under A you will see cells of Canadian pond weed. Under B you will see a one-celled animal called *Paramecium*. The *Paramecium* has been placed in water containing yeast cells which have been stained red. Look carefully at both slides, but take care not to move them. What is happening to the chloroplasts in the cells of Canadian pond weed? Can you explain how this is happening? Look for yeast cells inside the *Paramecium*. What is happening to these cells?

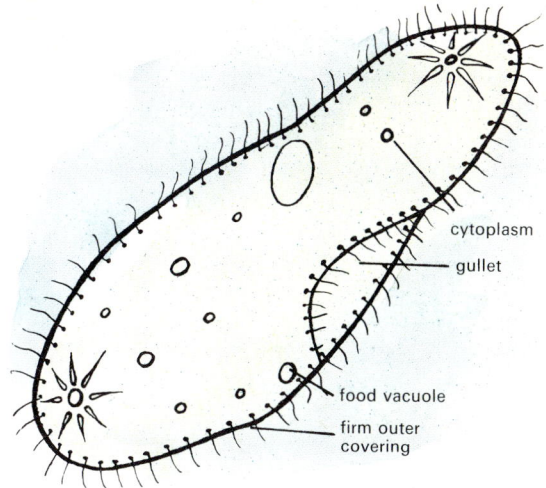

Fig. 14.21

In *Paramecium* food is taken in through the gullet. As the food is transported around the cell it is digested by enzymes. The digested food is absorbed into the cytoplasm. Undigested particles pass out through a pore which forms in the firm outer covering.

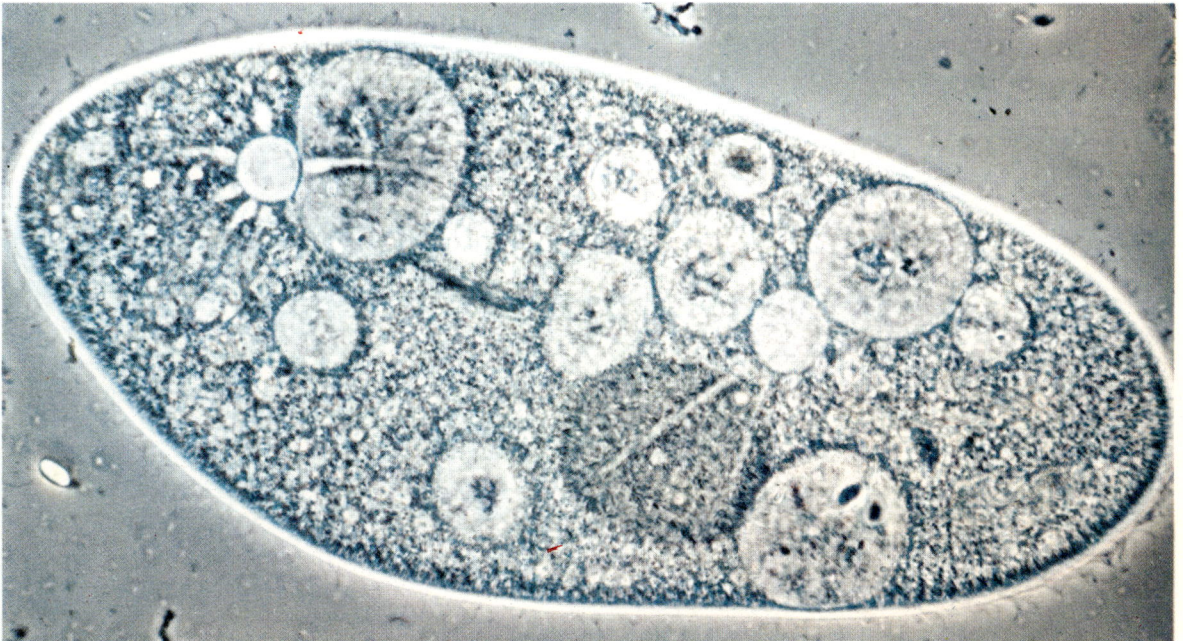

Fig. 14.22 *Paramecium.* Can you identify the various parts using Fig. 14.21?

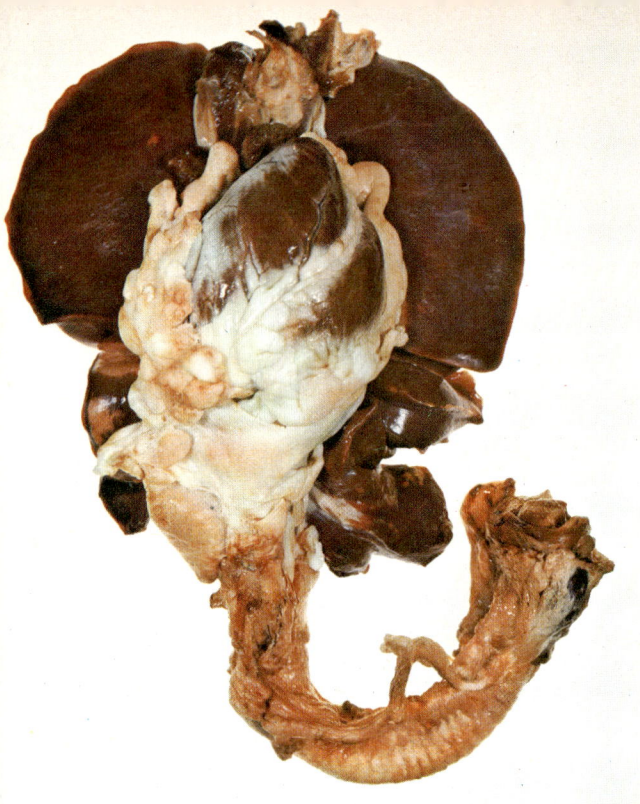

Fig. 14.23 Heart (surrounded by fatty tissue) and lungs of a sheep

fish out of water any longer than is necessary, read the following questions then look quickly down the microscope:

1. What colour is the liquid you see moving?
2. What do you think this liquid is?
3. Does it flow in tubes?
4. If tubes are present does the liquid flow in the same direction in all the tubes?
5. Are all the tubes the same width?

In a unicellular organism such as *Paramecium*, substances diffuse into and out of the cell. However as an organism increases in size, diffusion of substances from cell to cell is too slow a process to maintain life. A transport system and transporting liquid are necessary to carry substances from one part of the body to another.

In *Gammarus* this fluid travels throughout the body cavity. In goldfish the transport system is more highly developed. The transporting liquid, blood, flows through a series of tubes called blood vessels. Some of these blood vessels appear to take blood from the centre of the body to the extremities such as the tail, while others bring it back again. What brings about this movement of the blood? You will all know the answer to that question: it is the heart.

14.18 TRANSPORT SYSTEMS IN LARGER ANIMALS

Experiment 14.20

By means of a wide mouth pipette, transfer a small aquatic animal such as *Gammarus*, on to a watch glass. Observe it under the low power of the microscope. Look for movements of fluid within the body. Does the fluid appear to flow in tubes? Does this movement appear to be controlled in an organized way?

Experiment 14.21

Your teacher will place the tail of a living gold fish under the low power of the microscope. Do not be alarmed for the safety of the gold fish. If cotton wool soaked in pond water is wrapped around the head end, the fish can remain out of water for a short period without adverse effects. As you should not keep the

14.19 THE HEART AS A PUMP

Experiment 14.22

Look at a sheep's heart and the tubes which are attached to it. Press the top of the heart, then the bottom. Which end feels thicker? Your teacher will cut open the right side of the heart. How many cavities or chambers can you see? What do you see between the chambers? Are the walls of the chambers all the same thickness?

The upper chamber or **auricle** is a collecting chamber. Blood which has been round the body collects in it and then passes through a valve into the lower chamber, the **ventricle**. When the muscular ventricle contracts and pushes upwards the valve closes. Where does the blood go when this happens? Feel inside the right ventricle with a blunt seeker. You should find another hole at the top end of the ventricle. Push the seeker through the hole. Where does it go? You should find it coming out of the heart, through a tube. This tube takes blood to the lungs where gas exchange takes place. Blood rich in oxygen returns to the left side of the heart, to the left auricle. Look inside the left side of the heart. How many chambers do you see? Can you see a valve? This valve serves the same function as that separating the right auricle and ventricle. Find where blood leaves the left ventricle. Blood is pumped from the heart into **arteries**. It returns to the heart in **veins**.

Some veins run from the tip of your toes, up your legs and trunk to your heart. Measure the distance between your toes and your heart. Blood in these veins will have to travel that distance against gravity. What must be present in veins to stop the blood falling back? Similar structures prevented blood flowing from the ventricles to the auricles. What keeps the blood flowing in one direction in the arteries?

Experiment 14.23

Grip your bared arm tightly, just above the elbow, while your hand grips tightly on to a metre stick or a piece of wood. You will see lumps on the blood vessels. Can you suggest what these might be? Suggest a reason why they have formed.

The experiment you have just carried out was first performed many years ago by William Harvey. Find out when he lived and why his investigations on the circulatory system were so important.

Experiment 14.24

Find your pulse by placing two or three fingers of one hand (do not use your thumb) on to the opposite wrist. Count the number of beats in one minute. Write down your pulse rate, and then those of all your classmates. Draw a histogram of the results then answer the following questions:

1. What is the average pulse rate of your class?
2. Who has the slowest pulse rate?
3. Who has the fastest pulse rate?
4. Do the boys have faster pulse rates than the girls?

Measure 20 cm³ of lime water into a clean test-tube. Breathe *normally*, but bubble out each breath through a straw dipping into the lime water. Write down the number of breaths needed to turn the lime water obviously milky. Write down also your classmate's results. Draw a histogram of the results. What is the average for your class?

Experiment 14.25
What changes take place in the body as a result of exercise?

From the class, choose a group X of four or five pupils who will sit quietly for the rest of this series of experiments, but will also record their pulse rates and numbers of breaths.

Carry out some exercise — touch your toes and stretch up again ten times, or step on and off the top of the laboratory stool five times. If you decide to do the latter, ask someone to hold the stool firmly for you in case it topples over. Repeat the measurement of pulse rate and the number of breaths needed to turn 20 cm³ of lime water milky. List the results of the class and again find the average.

Carry out some vigorous exercise — step on and off the top of the stool thirty times, or run twice round the playground. Repeat the measurements and find the average.

Group X		Class groups	
Pulse rate	Number of breaths	Pulse rate	Number of breaths
Average			

Why were the results of group X recorded and used? What do we call this group? What happens to the pulse rate after vigorous exercise? What happens to the carbon dioxide content in the breath after vigorous exercise? Where did the energy required for these exercises come from? By what process is this energy liberated? Where does this process take place? What gas is a product of this process? Where does this gas leave the body? What transports this gas from one region of the body to another? Why did the heart beat faster after vigorous exercise?

You should now realize that in your body many complex processes take place. Digested food is carried by the blood to all the cells of the body. Some of this food is broken down by the process of respiration. The oxygen necessary for this reaction is carried by the blood from the lungs to the cells. Energy is released as a result of respiration. Carbon dioxide, a waste product, is carried by the blood from the cells to the lungs. During vigorous exercise, the body requires more oxygen for respiration and must get rid of the carbon dioxide. The heart therefore beats faster, and blood moves quickly round the body. Blood unites many systems in our body.

14.20 BLOOD

Experiment 14.26

Spin a sample of fresh blood in a centrifuge for five minutes, or examine blood which has been standing for one or two days.

Carefully remove the straw-coloured liquid from the top of the sample. This liquid is called **plasma**. Plasma carries food and carbon dioxide around the body. What foods would you expect to be carried by plasma: glucose, starch, protein, or all three? Give reasons for your answer.

Experiment 14.27

Plan an experiment to test the above hypothesis. Is your hypothesis correct?

The remaining blood in the centrifuge tube contains red blood cells and white blood cells. The red cells carry oxygen from the lungs to all cells in the body. White blood cells protect us by killing bacteria which enter our bodies.

14.21 BLOOD AND WASTE SUBSTANCES

An important role played by blood is the transport of waste substances. These are carried from the region where they are produced to the region which removes them from the body, or *eliminates* them. You know that carbon dioxide, a waste product of respiration, is carried from the cells of the body to the lungs. Carbon dioxide is only one of the many waste products produced in the body.

Excess proteins cannot be stored in the body because they contain substances which would poison us. They are broken down in the liver and the waste products formed are carried by the blood to the kidneys. The main breakdown product of proteins is urea, and this is excreted in the urine.

Experiment 14.28

Look at a dissected kidney and with the aid of Fig. 14.24 find the following regions: artery, vein, cortex, medulla, pelvis, ureter.

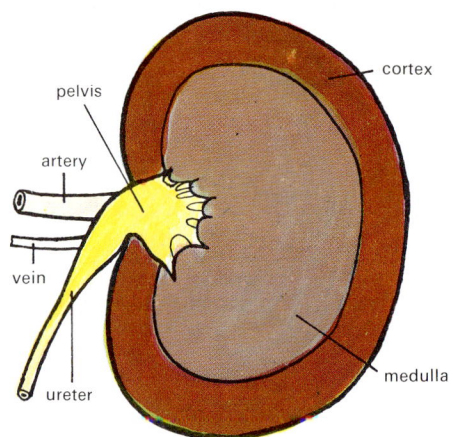

Fig. 14.24 Section through a kidney

Blood is carried by the artery into the kidney. Waste products and water pass from the blood vessels into the millions of fine tubes which make up the kidney, and blood free from impurities leaves via the vein. The tubes in the kidney unite to form the ureter. Waste products and water, which together form **urine**, pass from the tubes to the ureter and then to the **bladder** from where they are eliminated. The kidneys are said to **excrete** urine.

In the description above, water is mentioned as a waste product, and yet water is essential for us to live. You can live without food for weeks, but only a few days without water. Why then does our body eliminate water? The answer is simple. We usually drink more water than we need, and we must get rid of it or we would burst.

14.22 GETTING RID OF WATER

Experiment 14.29

By measuring the volume of common house-hold containers, e.g. tea cups, tumblers, milk bottles, find the volume of liquid you drink in the course of a normal day. What weight of water do you take in each day? (One litre of water weighs 1 kg.) If each urination gets rid of about 200 g of water, what weight of water do you lose by this method each day? What weight of water would appear to be still left in your body each day?

Does your body weight increase by this amount each day? If it does not, how do you get rid of the rest of the water? The following experiment will help you to answer this.

Experiment 14.30

Cut about twelve pieces of filter paper so that each measures 2 cm². Place each piece on a strip of adhesive tape so that the tape extends beyond the filter paper. Weigh all the strips together. Attach the strips to different regions of the body, e.g. on the wrist, the inside of the elbow, under the arm, so that the filter paper is next to the skin. Wrap a piece of polythene over each strip (if this is possible). Leave the paper in position for about an hour, and do some exercise during that time. You could select one pupil as a 'guinea pig' and send him a few times around the playground. Remove the adhesive tape and filter paper and reweigh. Can you account for the difference in weight?

14.23 SWEAT

You lose a large volume of water each day as sweat. Sweat evaporates from the skin at all times, but the amount of water lost in this way depends on our behaviour. When do you lose most sweat? The following experiment will help you to understand why this happens.

Experiment 14.31

Place a drop of ether on the back of your hand. How long does it take for the ether to evaporate? Does your skin feel hot or cold?

The rapid evaporation of ether cooled your skin. When sweat evaporates, heat is absorbed from the skin and the body temperature drops. Sweat is therefore important in controlling body temperature. Why do you sweat profusely when you are active? When will little sweat be produced by the body?

Water leaves the body as sweat, in urine and in your breath. When do you see the latter? If little water is lost in sweat, the amount of water in urine will increase. The converse is also true. If you lose a great deal of water through sweat you feel thirsty, and drink to replace the lost water. In this way the amount of water in the body is kept fairly constant.

14.24 YET ANOTHER WASTE PRODUCT

Earlier in this unit you learned that food is digested in the alimentary canal, and that the digested foods are absorbed into the blood vessels in the intestine wall. However, not all the foods we eat are digested. Some parts of our food cannot be digested because we have no enzymes in our bodies that can digest them.

The food material that passes into the colon after digestion is completed is therefore waste and must be removed from the body. The colon contains large quantities of bacteria. Some of these bacteria, together with undigested food pass at regular intervals from the anus as **faeces**.

Make a summary of the removal of waste products in man by completing the following table:

Waste product	Region from which waste produced is removed
1	
2	
3	
4	

14.25 WASTE PRODUCTS IN PLANTS

In animals, waste products are removed from the body by definite excretory systems. Such systems are not found in plants. This does not mean that plants do not produce waste products. They do, but the amount of waste produced is much less than that produced by animals. Can you suggest why this is so? Waste products in plants are transported to parts of the plant that will be shed, e.g. leaves, petals, fruits, and bark.

The waste products occur in a variety of forms. Some are found as oils and they produce odours in leaves. Eucalyptus, lavender, and clove oil are all waste products. Others are deposited as crystals in leaves and fruits. When the leaves and fruits fall from the trees, the waste products are removed. The gritty pieces in a 'gritty' pear are crystals of waste material. Most waste products are removed from deciduous trees in autumn. As evergreens shed their leaves regularly throughout the year, waste products are removed continuously.

Questions for you to answer

1. What gas is produced as a waste product in plants during the day?
2. What gas is produced as waste product by plants at night? Account for the difference.

WHAT YOU HAVE LEARNT IN THIS UNIT

1. Food is made up of carbohydrates, proteins, and fats, together with small amounts of mineral salts and vitamins. Animals also need water, but it cannot be classed as a food as it does not contribute either to growth (or cell replacement) or energy.

2. Starch and sugars (such as glucose and cane and beet sugar) are carbohydrates.

3. Energy is obtained mainly from carbohydrates and fats; proteins are used for growth and cell replacement. Energy is released in a process known as respiration in which the carbohydrate or fat is 'burnt up' through the action of oxygen in the air we breathe. The process is very complicated, although the final effect is the same as if the food had been burnt.

4. For proper health and growth it is necessary to have a balanced diet containing the correct proportions of carbohydrate, protein, and fat.

5. The process of digestion of food is the breaking down of the complex molecules which make up food into simpler ones which can pass through the walls of the intestines into the blood stream. The digestion of food begins in the mouth. Many animals have teeth with which food is shredded and broken into small pieces. It is mixed with saliva, which contains enzymes which start the process of breaking down the food. It is continued in the stomach and the small intestine. Enzymes are poured on to the food from glands connected with the alimentary canal.

6. Tooth decay is caused by acid produced by the action of bacteria on particles of food trapped between the teeth. It can be prevented by removing the food particles by brushing, and by strengthening the teeth by the addition of fluoride to water.

7. The teeth of animals are adapted to their modes of feeding. The dentition of the sheep, for instance, is adapted to chewing; that of the dog to tearing up meat.

8. The blood transports digested food to different parts of the body in the blood plasma. It also transports oxygen from the lungs, and waste products to parts of the body from which they can be eliminated.

9. The blood is pumped through the body by the heart. It leaves the heart by the arteries, and returns to it by the veins.

10. Waste products include carbon dioxide gas which is eliminated through the lungs, water which is excreted as urine and as sweat, and solid waste which consists of undigested food and which is excreted from the anus. The body cannot store excess protein. It is broken down in the liver and the waste from this is carried to the kidneys by the blood. The kidneys excrete urea in the urine; urea is the breakdown product of proteins.

11. The waste products of plants are transported to parts of the plant which will be shed, e.g. the leaves, petals, fruits, and bark. Sometimes these waste products are useful to us, e.g. substances like oils – lavender, clove, eucalyptus – and sweet-smelling products in petals from which perfumes are made.

Unit 15
Electricity and Magnetism

15.1 SAFETY AND DANGER WITH ELECTRICITY

You have already done quite a number of experiments with electricity earlier on in this course (Unit 7, Book 1), and in this unit we are going to find out how we can use electricity safely at home and at work.

Electricity at the mains voltage can be dangerous if we do not use it properly, and we must be sure that we know how to handle electrical appliances without getting electric shocks. We must first remind ourselves about those materials which conduct electricity and those which do not. You will remember that the former are called conductors, and the latter insulators.

Conductors	Insulators

15.2 ELECTRICITY IN THE HOME: HOW APPLIANCES ARE CONNECTED TOGETHER

Experiment 15.2

Fit up on your circuit board the components listed in Fig. 15.2. You have already studied a circuit rather like this in Unit 7 (Book 1, p. 114). What kind of circuit is it?

Experiment 15.1
Conductors and insulators

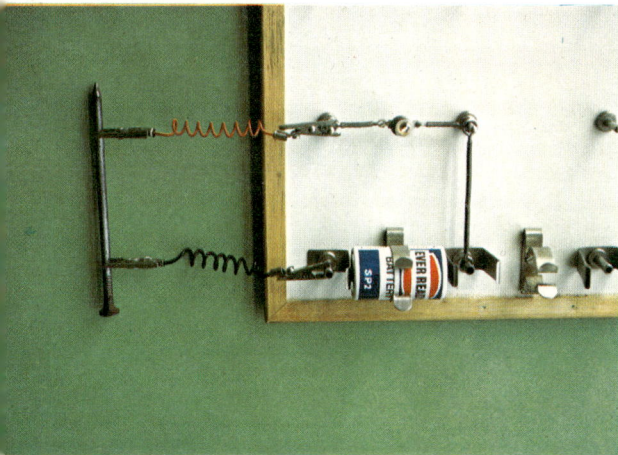

Fig. 15.1

Fit up the circuit shown in Fig. 15.1 and use it to find out whether materials are conductors or insulators. Connect the materials under test in place of the nail. What will tell you if the substance is a conductor or an insulator?

Make lists of the materials that you test under these two headings, and against each say what use they have in electrical appliances.

Fig. 15.2

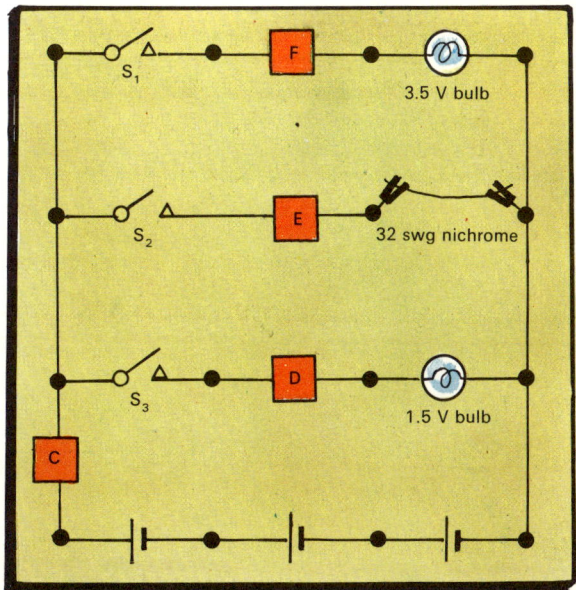

What is the advantage of circuits like this? We can soon see. Try the effect of closing the switches one at a time for a few seconds, and then in whatever combinations you like. You will find that the bulbs and the thin wire can each be switched on or off independently of the others.

The length of thin wire should heat up a little when switch number 2 is closed. The bulbs and the thin wire represent different appliances you might have in a house. Can you suggest what each could represent?

Now, with all three switches closed, replace links C, D, E, and F by an ammeter, and write down the readings you find in each case. If you add all the currents at D, E, and F together, what do you notice about their sum compared with the current at C?

Position of ammeter	Current (A)
C	
D	
E	
F	

The current at C is the total current flowing out from the cells. This branches into the different paths, D, E, and F, and as there can be no loss or build up of charges in the circuit, current C must be equal to the sum of the other three currents.

Do you know where the wires bringing electrical energy to your house enter the building? The total electric current used in your house at any particular time flows through these wires; it then branches into all the electric circuits you have for the appliances you are using. The total electric current is like that at C in your experiment; the smaller currents passing through whatever you may be using – lamps, TV set, heaters, electric iron, hair drier, and so on – are represented by the currents at D, E, and F in your experiment.

15.3 MEASURING ELECTRICAL ENERGY

How do the Electricity Board officials know how much energy is used up, so that they can send you the right bill for your electricity? You will have an electricity meter at home which measures the

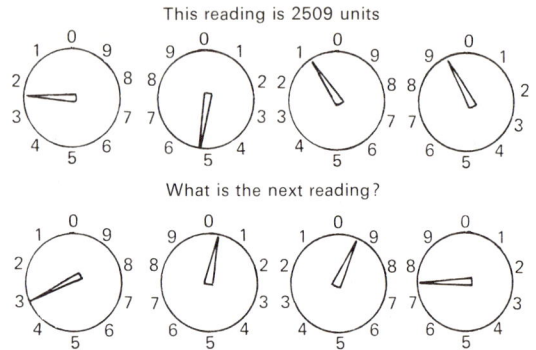

This reading is 2509 units

What is the next reading?

How many units have been used between meter readings?

Fig. 15.3

amount of energy you have used. When current flows through the meter a disc rotates, rather like an electric motor. The more current flowing, the faster the disc turns. It is connected up through a train of gears with hands which move round circular scales. Look at the electricity meter at home, and ask someone to switch on various appliances. Notice how fast the disc rotates. Does it turn faster when an electric fire is switched on than it does when there is only a single lamp on?

Experiment 15.3
Experiments with an electricity meter

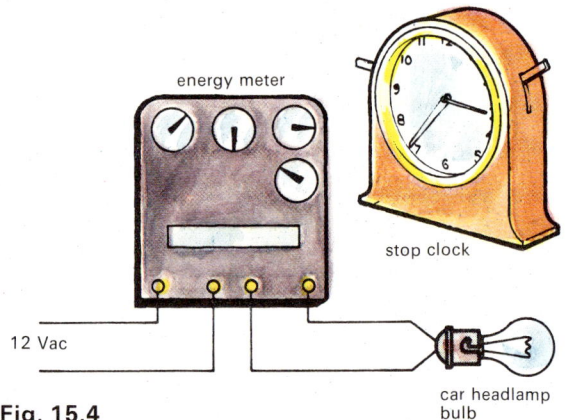

energy meter

stop clock

12 Vac

car headlamp bulb

Fig. 15.4

Connect up some 12 volt bulbs rated at 6 watt, 24 watt, and 48 watt with a power pack and an electricity meter (called a Joulemeter, because it records the electrical energy used in joules). To read the scales remember to take the number the hand has just left. The hands on the dials go alternately clockwise and anticlockwise; this can be confusing until you get the hang of it. (Some electricity meters nowadays do not have dials but have a direct reading indicator instead, very similar to the mileage indicator on a car. They are much easier to read than the meters with dials.)

Read the meter and switch on each lamp for 1 minute 40 seconds. After each has been on for this time (which you will see is 100 seconds) read the meter again. The difference between the two readings gives you the number of joules of energy the lamp has used in 100 seconds. Calculate how many joules it has used in 1 second. Do you remember another name for a joule per second? Compare the number of joules per second in each case with the power of the lamp in watts. What do you find?

Power really means the rate at which energy is transformed, and what you have just done should remind you that the power of the appliances in watts is simply the number of joules of energy each uses in 1 second. Another way of putting this is that 1 joule of energy is used by an appliance of power 1 watt in 1 second, i.e.

$$\text{Power (in watts)} = \frac{\text{Energy (in joules)}}{\text{Time (in seconds)}}$$

or

$$\text{Energy (in joules)} = \text{Power (in watts)} \times \text{Time (in seconds)}$$

Check these rules with the results you have just obtained.

At home, with more powerful appliances and bigger voltages, a bigger unit of energy than the joule is required; just as if we wanted to measure a long distance, such as that between Edinburgh and Glasgow, we should not measure it in metres, but in larger units, like the kilometre. This bigger unit of electricity is the quantity of energy that an appliance of power 1000 watt (or 1 kilowatt) would transform in 1 hour. It is called the kilo-watt-hour (abbreviated to kWh), and is often just called a UNIT of electricity – short for a Board of Trade Unit. Calculate how many joules one kWh is equivalent to. To do this multiply the number of watts in 1 kW by the number of seconds in 1 hour.

15.4 PAYING FOR ELECTRICAL ENERGY

To find the number of units (or kilowatt hours) of electrical energy used we multiply the power in kilowatts by the time in hours for which it is switched on.

$$\text{Number of units (kWh)} = \text{Power (in kW)} \times \text{Time (in hours)}$$

Thus in 3 hours, a 2 kW heater would use 6 units. The cost of each unit you use at home will vary between 3·5p and 1p, but on the average it will be about 1·5p.

To get some idea of typical costs, work out the following, supposing electricity to cost 1p per unit.

(a) The cost of using a 3 kW heater for 10 hours.
(b) The cost of using a 150 watt TV set for 4 hours each day this month.
(c) The cost of using a 5 watt electric clock, 24 hours a day, for a year (365 days).

15.5 CURRENTS AND FUSE RATINGS

How large a current flows through a lamp, or an electric heater, or a television set? Let's find out. Because it is dangerous to work with the electricity mains your teacher will set up this experiment, but you should take the readings of the ammeter. *You must not touch any part of the apparatus.*

Experiment 15.4

The apparatus shown in Fig. 15.5 will be set up by your teacher. Take the reading of the voltmeter and the ammeter when a 60 watt lamp is in the circuit. Then repeat with lamps of other powers, such as 75 watt and 100 watt. What do you notice about the product of the voltage in volts, and the current in amps in each of these cases? How does it compare

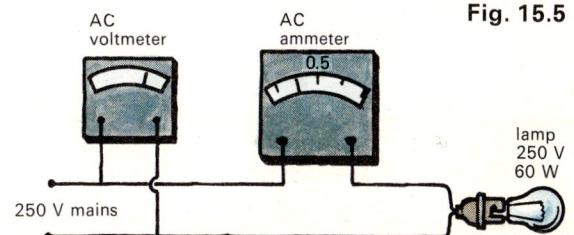

AC voltmeter

AC ammeter

Fig. 15.5

250 V mains

lamp 250 V 60 W

Appliance	Power (watts)	Voltage (V)	Current (A)	Correct fuse (1,2,5,13 A)
Immersion heater		250 V	13 A	
Hot plate	2 kW	250 V		
Electric iron		250 V	3 A	
TV (U.K.)	150 W	250 V		
TV (U.S.A.)		110 V	1.4 A	
Radio		250 V	0.2 A	
Light bulb	100 W	250 V		
Car headlamp		12 V	4 A	

with the wattage of the lamp? We find that

Power (in watts) = Voltage (in volts) × Current (in amps)

so that

$$\text{Current (in amps)} = \frac{\text{Power (in watts)}}{\text{Voltage (in volts)}}$$

Use these rules to complete the table of values given above.

You will remember from Book 1 (page 117) that fuses are put into electrical circuits so that they will break the circuit by melting if the current is too great. The fuse values should be those just bigger than the current the appliance requires. Thus, if you had a 2400 watt electric fire running on a voltage of 240 volts, it would pass a current of 10 amp. If the fuse you inserted in the circuit was a 5 amp one, it would melt every time you switched the fire on. It is necessary to put a 13 amp fuse into this circuit. This gives you a little margin, and it would certainly melt and break the circuit if anything went wrong, such as two bare wires touching.

15.6 ELECTRONICS, OR ELECTRONS IN MOTION

In Unit 7 (Book 1) you found that some substances (insulators) could be charged by rubbing them. What charges are usually given in this way to polystyrene (the opaque plastic) and to cellulose acetate (the clear plastic) strips you used?

Experiment 15.5
Charges

Take one of your charged strips and touch it against the cap of an electroscope (Fig. 15.6). What do you see happening to the gold leaves? Why do you think this happens?

If you have used a charged cellulose acetate strip you will have put positive charges on the cap of the electroscope. These travel down to the leaves, and as both of them have positive charges they will repel each other. Now remove the plastic strip and touch the cap of the electroscope with your finger. What happens to the leaves? Why does this happen? Are the positive charges still there? What kind of charges would be needed to cancel them out?

These negative charges must have come from somewhere, possibly from your body. In any case they must have travelled *through* your body to get there. The negative charges, you will remember, are called **electrons**.

There are several other ways of discharging an electroscope. Try bringing a Bunsen flame near the cap of a charged electroscope. What happens? Where, do you think, have the electrons come from that are necessary to cancel out the positive charges?

It is worth while trying out the experiments with the van de Graaff generator that you did in Unit 7 (Book 1) again just now in order to remind yourself of them. You will remember that in one experiment we had wooden leaves fixed to the dome of the generator. When the dome is charged positively one leaf rises if the air is dry enough. Here is an opportunity to try different methods of discharging the dome.

Fig. 15.6

gold leaves of electroscope

Experiment 15.6
Cancelling out charges

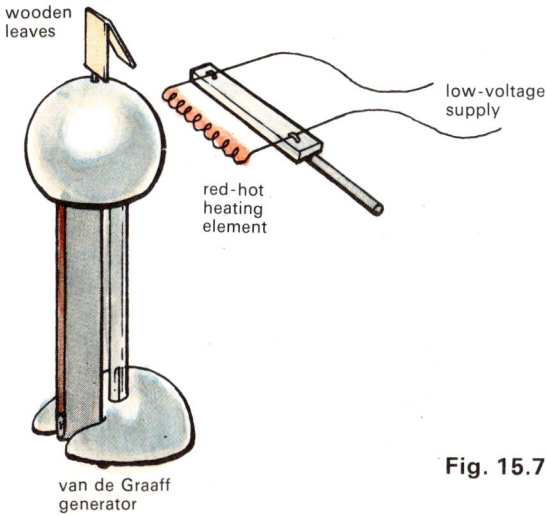

wooden leaves

red-hot heating element

low-voltage supply

van de Graaff generator

Fig. 15.7

Find out what happens when a heated spiral of wire is held near the dome (Fig. 15.7). What must the heated wire be giving off in order to cancel out the positive charges on the dome?

These electrons can only have come from the atoms of the wire. Perhaps when the wire is heated enough the atoms in it vibrate so much that the electrons on the outside of some atoms are flung off and are attracted by the force due to the positive charges on the dome. The voltage of the dome may be as much as 250 000 volts so it will certainly have a strong attractive force for electrons.

15.7 ONE WAY CONDUCTION

Here is another experiment your teacher will demonstrate.

Experiment 15.7

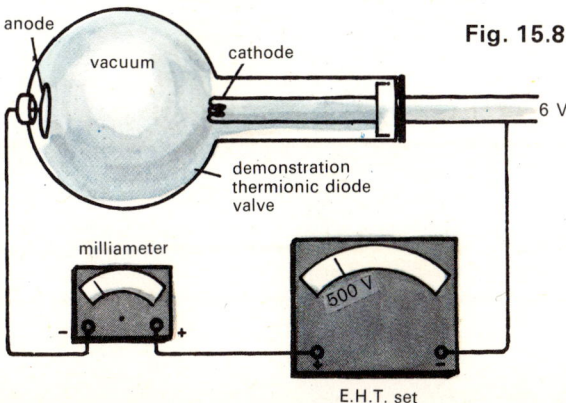

anode

vacuum

cathode

6 V

demonstration thermionic diode valve

milliameter

500 V

E.H.T. set

Fig. 15.8

The circuit shown in Fig. 15.8 is fitted up. What happens to the spiral of wire (called the filament) in the evacuated tube when it is connected to a 6 V supply? It gets hot. What particles will it be giving off? What kind of charge do they have?

When the extra high tension (E.H.T.) supply is switched on, positive charges will be put on to the disc (called the anode) in the tube. Watch what happens to the milliammeter. If it shows a reading you know that a stream of electrons must be passing through it. Trace out the complete circuit in which the electrons are flowing. They must flow from the filament (sometimes called the cathode) through the empty space in the tube to the anode, through the wire to the meter, then to the positive terminal of the E.H.T. unit and out from the negative terminal to the cathode.

The connection on the E.H.T. unit will now be reversed so that there is a negative charge on the anode. Is there any current flowing through the meter now? Explain what is happening. Remember that like charges repel each other, so that when we put negative charges on the anode these will repel the electrons driven off from the cathode.

Obviously, electrons will flow through this tube in one direction only. They will only flow when the anode is given a positive charge and not when it has a negative charge. What do we call devices which allow flow in one direction only? You have something like this controlling flow of air into your bicycle tyres; it will allow air to flow in but not out. Similarly we have this kind of thing preventing the blood in our veins from flowing backwards.

The tube we have been using in the last experiment is a kind of electronic valve. Because it has two parts, an anode and a cathode, it is called a **diode** valve. Diode valves are used to convert alternating current (i.e. current which flows first one way and then the opposite way) into direct current (i.e. current which flows only one way). Your household electricity supply is alternating. It changes direction 100 times a second. On the other hand, the current obtained from a battery is direct current. It flows only in one way. A stream of electrons flows from the negative terminal to the positive one. When an alternating current is converted into a direct one it is said to have been **rectified**. The diode valve can be used as a rectifier. Television sets and mains radio sets contain rectifiers to convert the household alternating current into direct current required for the proper working of the set.

The electrons given off by a hot wire are sometimes called thermions, and this kind of valve is therefore called a thermionic valve. Nowadays this kind of valve is often replaced by a transistor.

15.8 TRANSISTOR DIODES

Experiment 15.8

Fig. 15.9

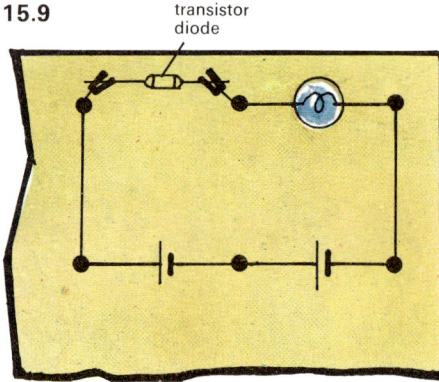

transistor diode

Set up the circuit shown in Fig. 15.9 on your circuit board. What happens to the bulb when you take out the transistor and reverse it in the circuit? Replace the transistor by a resistor marked with one brown and two black bands. When you reverse the resistor in the circuit, does this have any effect on the lamp?

We come to the conclusion then, that current can flow in either direction through a resistor, but in one direction only through a transistor.

Experiment 15.9
The Maltese Cross Tube

So that you can understand even more clearly what happens in a diode valve, your teacher will set up a circuit using a Maltese Cross Tube, as shown in Fig. 15.10. This tube is very

Fig. 15.10

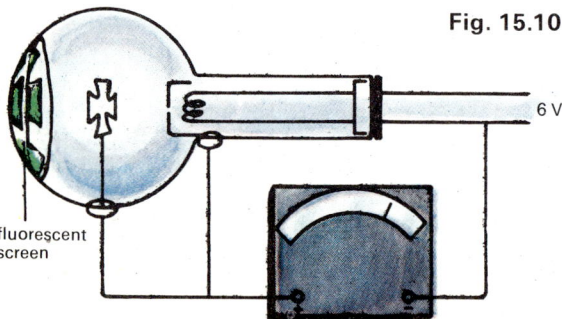

fluorescent screen

6 V

much like a diode. The filament is the cathode and this time the anode is a cylinder. Do you remember what kind of charge must be given to the anode if it is to attract electrons away from the cathode? The cathode is often called an 'electron gun' because electrons are shot off from it.

One end of the tube is painted with a substance which glows when electrons hit it. This is called the screen. Watch what happens to the screen when positive charges are put on to the cross. You will see a shadow on it. What does the shadow look like? We can explain this effect by supposing that electrons shot off from the cathode fly through the anode. If they are stopped by the Maltese Cross they cannot reach the screen, which therefore remains black at this point. If, however, they do get by — and they will do this if they do not hit the obstacle — they make the screen glow.

What happens when a strong magnet is brought near the tube? What must be happening to the beam of electrons? Can you suggest another way of getting much the same effect?

Sometimes the electrons flying off the cathode are called 'cathode rays' because the electrons come from the cathode, and they speed along the tube in a straight line, like rays of light. A special version of this kind of tube is used in the cathode ray oscilloscope (c.r.o.) which you have already used in Unit I and in a TV tube.

15.9 THE CATHODE RAY OSCILLOSCOPE

Experiment 15.10

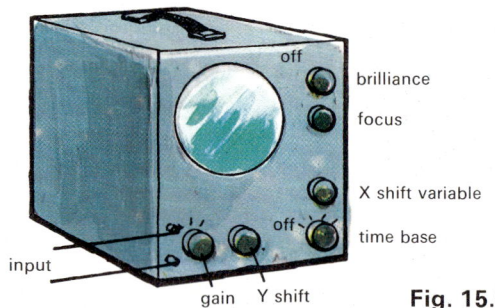

off — brilliance
focus
X shift variable
off — time base
input
gain   Y shift   **Fig. 15.11**

Switch on the oscilloscope. You will have to wait a few seconds before seeing anything on the screen. For our purpose it is best to start with the 'time base' switched off. If after about

10 seconds there is still nothing to be seen, try turning the 'shift' and 'variable' knobs in turn, and you should find that on turning one of them a blob of light appears on the screen. The 'shift' control will move up or down, and the 'variable' knob will move it from left to right. Try to bring the spot to the centre of the screen. Now adjust the 'brightness' and 'focus' knobs to bring the blob of light to a sharp point and a suitable brightness. What causes the blob of light? Try bringing up a magnet to the screen. What happens? Can you suggest how the 'shift' and 'variable' controls might work?

The tube in the c.r.o. is very like that used in the last experiment and has at the back an electron gun firing electrons through a hollow anode on to the screen. Switch on the time base to positions 1, 2, and 3 in turn and watch what happens to the spot travelling across the screen.

In a television tube the electrons move across the screen in lines which are very close together – 405 or 625 lines in the width of the screen. Which do you think will give the better picture, 405 or 625 lines?

Connect the output from a small generator (alternating current) to your c.r.o. What do you see on the screen? What happens when the generator is turned faster? Connect a microphone to the c.r.o. input terminals, and sound a tuning fork in front of it. Can you explain the pattern you see on the screen? Sing or speak into the microphone. What kind of pattern do you get this time? Why is it more complex than the pattern you get with a tuning fork?

15.10 STRIP LIGHTING

Experiment 15.11

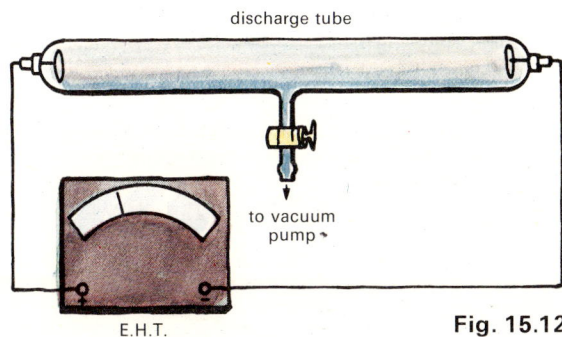

Fig. 15.12

The apparatus needed for this experiment is a long glass tube containing two plates or wires connected to terminals at the end of the tube. The tube can be connected to an air pump which will suck the air out of it. A voltage of 5000 volts is connected across the two terminals from an E.H.T. supply. Does anything happen?

The pump is now started up. Watch what happens as the air is removed. When the pressure gets low enough you will see a violet glow in the tube; this splits up into bands which get further and further apart as the air is taken out of the tube.

The charge on the positive anode is sufficient to pull out electrons from the metal atoms of the negative cathode. If, however, there is air in the tube at ordinary pressure these electrons do not get very far before they collide with a molecule of oxygen or nitrogen. When the air particles are removed by the pump so that there are comparatively few of them left, the electrons can travel quite a way before they hit one. If the electrons are able to avoid collision with air particles at first, they speed up so much that when they do collide with air particles they have sufficient energy of movement to cause the air particles to glow. This principle is used in the fluorescent tubes which are often used nowadays for electric lighting and for advertisement signs. In these tubes, not only does the gas in the tube give out visible light, when it is bombarded by electrons in this way, but invisible ultra-violet light as well. This ultra-violet light can cause substances painted on to the tube to glow.

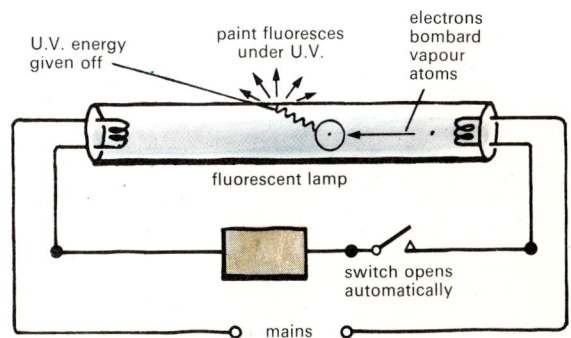

Fig. 15.13

Fluorescent tubes, being long, cast very little shadow and are very efficient at transforming electrical energy into light energy. The other type of electrical lamp, as you know, has a thin spiral of wire inside it (called the filament) which gets hot when an electric current is passed through it. It is not so efficient as the fluorescent tube because much of the electrical energy is converted into heat (which is often not required), and only some of it is converted into light energy. A 40 watt fluorescent tube can give out as much light as a 100 watt filament lamp.

The colour of the light given out by a fluorescent tube is governed by the kind of gas in the tube and by the substance used to coat the interior. By choosing these correctly the light given by the tube can be made very nearly the same as daylight. The orange-coloured light given out by some street lamps is produced by tubes which have sodium vapour in them – you already know, from your work in Unit 12, page 52, that sodium atoms when heated give out a yellow-coloured light. The street lamps which give out a bluish-white light contain mercury vapour. Red, green, and blue advertising signs contain the noble gases.

15.11 ELECTROMAGNETISM

You have already encountered those fascinating objects called magnets earlier in this science course. Do you remember the rings which seemed to be able to float on top of each other (Book 1, page 15)? What happened when one of the rings was reversed? If you are not sure, try the experiment again. Magnets are able to attract certain materials. What are they?

We now want to investigate one of the very surprising results of passing an electric current through a wire.

Experiment 15.12

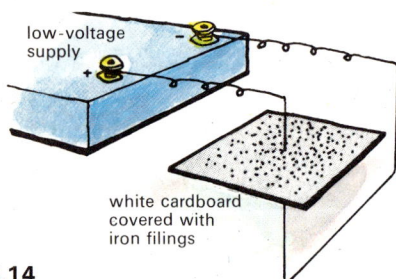

low-voltage
supply

white cardboard
covered with
iron filings

Fig. 15.14

Fit up the apparatus shown in Fig. 15.14, using a length of 25 cm of copper wire threaded through the hole in a piece of white cardboard. Connect the ends to the red and black terminals

of the low-voltage supply. Do not switch on yet. If this is a *special low-voltage, high-current power pack* the length of time the current flows is not important, but if you are using an accumulator or dry battery to supply the current, it must not be left on for more than about 5 seconds at a time.

With the cardboard horizontal, lightly sprinkle some iron filings on to it. Switch on the current, and tap the cardboard lightly with a pencil. What happens to the filings? Switch off, and draw the pattern you see.

You will remember that iron was one of the substances that was attracted to a magnet. Was copper one of the materials you found to be magnetic? Can you suggest a reason for the iron filings arranging themselves in circles round the wire when the current was flowing? Does this happen when the current is not flowing?

Put the iron filings back into the shaker, and now put a small plotting compass on the cardboard and switch on the current again. What happens to the compass needle? Mark the positions of the ends of the compass needle with a pencil, switch off the current, and draw a line joining the two pencil dots, adding the arrowhead to the line marking the direction in which the needle pointed. The arrow end of the compass needle is the end which normally points to the north, and is called the north (or, more correctly, the north-seeking) pole of the needle. Now move the needle to the end of your line, and repeat the experiment. Do this several times until you have moved the compass needle right round the wire. Is there any connection between the effect on the iron filings and that on the compass needle?

You will find that the compass needle points along the lines on which previously the iron filings lay. We call these lines 'lines of force'. Using your compass needle as your guide, say whether they point clockwise or anticlockwise round the wire. Now reverse the supply connections to the wire and repeat the process. Do the lines run in the same direction as before?

The pattern of lines of force is called the 'magnetic field'. In science a 'field' is anywhere where an effect can be felt. Thus, the field of a magnet (or a magnetic field) is the area in which the force of the magnet can be detected. The gravitational field of the earth is the space round the earth where the force of gravity acts. The electrical field of a charged body is the space round it where the charge can exert an influence. In a magnetic field the lines of force act in the direction in which the north pole of the plotting compass needle points. There is a simple rule which will tell you the direction of the concentric circular lines of force which surround a wire carrying a current; if you point your thumb in the direction of what we call the conventional electric current in the wire (i.e. supposing the current to flow from the + to the − end of the wire, just opposite to the actual direction in which the electrons flow) and bend your fingers round the wire, the direction in which your fingers curl is the direction of the lines of force (Fig. 15.15). Try it and see if this rule is correct.

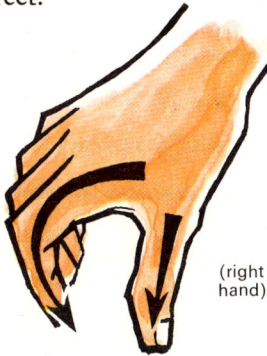

(right hand)

Fig. 15.15

Experiment 15.13
The magnetic field of a coil

Wind a coil of about ten turns of copper wire (plastic covered) on a wooden cylinder, and then insert this coil over the cardboard as shown in Fig. 15.16. Connect up the coil to a special low-voltage power pack, or to an accumulator or dry battery, but do not switch

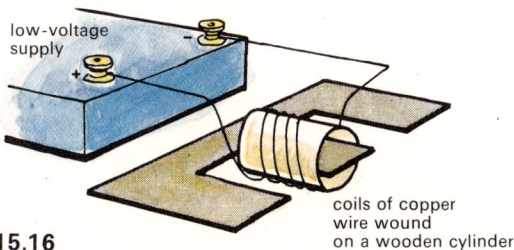

low-voltage supply

coils of copper wire wound on a wooden cylinder

Fig. 15.16

on. First find out if the coil has any effect on a plotting compass needle when there is no current flowing. Now switch on, and put the plotting compass at each end of the coil in turn. What happens? If one end of the coil attracts the north pole of the needle what does this mean?

You will find that when the current flows in the coil, one end of the coil becomes a south pole and the other a north pole. Do these poles remain when the current is switched off? Repeat the experiment with the current reversed. Do you get the same poles at the ends of the coil, or are they reversed too? What do you think would happen if you passed alternating current through the coil?

Now repeat the experiment, this time sprinkling the cardboard lightly with iron filings. When you switch on the current, tap the cardboard lightly with a pencil. Switch off, and draw the pattern of lines of force you get.

You might also try this experiment using a longer wire and coiling it into about thirty turns. Does this have any effect on the pattern of lines of force or on the way the compass needle is attracted?

It is very useful to know some rule for finding out which sort of pole we have at the end of a coil when the current is flowing in a certain direction. Go back to the coil and look at one end of it. In which direction does the conventional current flow from your point of view? What pole of the compass needle is attracted? You should find that if the conventional current is flowing clockwise it is the north pole that is attracted; in other words the end of the coil acts as a south pole.

Similarly the end of the coil in which the current is flowing anti-clockwise behaves like a north pole. From the diagram you will see that this is very easy to remember (Fig. 15.17).

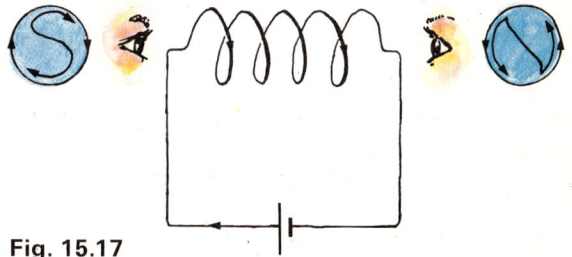

Fig. 15.17

The correct pole is that whose first letter, when arrows are attached to the ends of it, shows the direction of the conventional current flow as you look at the coil.

15.12 THE MAGNETIC FIELD OF A BAR MAGNET

It is interesting to compare the pattern of the lines of force set up when a current flows through a coil of wire with that from a bar magnet.

Experiment 15.14

Place a piece of thin card or stiff paper over a bar magnet and sprinkle iron filings on the card. Tap the card lightly with a pencil. (Fig. 15.18). You will find, rather surprisingly, that

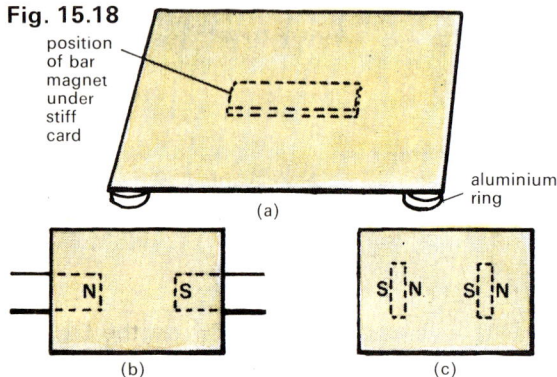

Fig. 15.18

position of bar magnet under stiff card

aluminium ring

(a)

(b)   (c)

the pattern is identical with that for a coil of wire carrying a current. Notice that in both cases the lines of force appear to be curves which seem to start from the north pole and run round to the south pole, in no case cutting through any other lines of force.

Experiment 15.15

It is important to find out the pattern of lines of force when two magnets are placed about 5 cm apart with north and south poles facing towards each other. Remember these are the poles which attract each other.

This time the lines of force run between the magnets from the north pole of one to the south pole of the other, just like tight elastic bands.

15.13 ELECTROMAGNETS

We now want to find out if coiling our wire round a piece of iron makes any difference to the magnetism produced when the current flows.

Experiment 15.16

Fit up the apparatus in Fig. 15.19 using a coil of about twenty turns. Now place a card on top of the iron C-core, or iron bar, and sprinkle some iron filings on the card. Switch on the current and map the field as before. From the

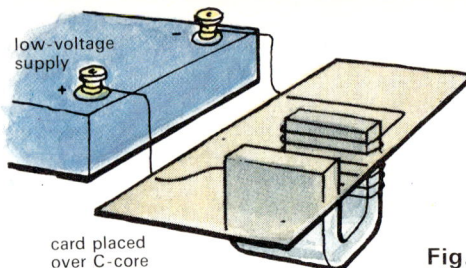

low-voltage supply

card placed over C-core   Fig. 15.19

effect on the iron filings do you think the iron core has made any difference to the magnetism produced?

What you have made in this experiment is called an 'electromagnet'. Try the effect of bringing up another C-core when the current is flowing. Try to separate them once they are joined. Now switch off the current. Do you find the same difficulty in separating the C-cores? Try to find out some uses for electromagnets in hospitals, shipyards, scrapyards, and so on.

Electromagnets are important parts of telephones, electric bells, motor-car horns, and hosts of other things. See if you can spot them in any examples of these appliances you may have in the laboratory.

15.14 THE FORCE ON A CONDUCTOR CARRYING A CURRENT IN A MAGNETIC FIELD

We are now going to find out what happens when we pass a current through a wire which is placed in a magnetic field which crosses the wire so that the lines of force are at right angles to the wire.

Experiment 15.17

Fit up the apparatus shown in Fig. 15.20 using three pieces of bare copper wire each 8 cm long. Switch on the current and bring up the magnets

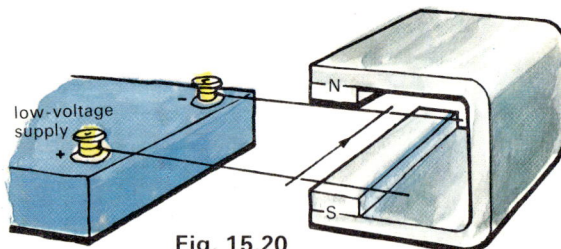

low-voltage supply

Fig. 15.20

as shown in the diagram. What happens?

Now repeat the experiment, this time inverting the magnet. You should find that in one case the third bare wire lying across the other two parallel wires is forced to move to the left, and, in the other, to the right.

Experiment 15.18

Here is another way of showing that a wire carrying a current is acted upon by a force when it is placed in a magnetic field. Set up the arrangement shown in Fig. 15.21. When the current

Fig. 15.21

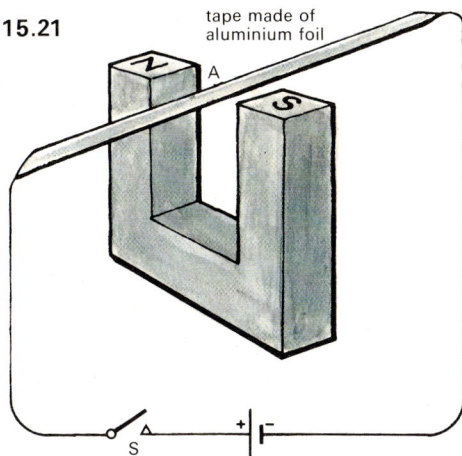
tape made of aluminium foil

is switched on, what happens to the foil? Reverse the magnet. Is there any difference when you switch on the current?

Are the results of these experiments what you would have expected? After all, when a current is passed through a wire it produces a magnetic field. You know, too that there are forces between magnetic fields; magnetic poles either attract or repel one another. It is not surprising, then, that there will be a force between a magnet and a wire carrying a current. This force is a very important one, and its existence makes life easier for us all. It is the force which operates in electric motors.

15.15 PREDICTING THE DIRECTION OF THE FORCE

As we know the direction of the magnetic field produced when a current flows in a wire, it ought to be possible for us to work out the direction of the force which acts on the wire when it is placed in a magnetic field. There is a simple rule which will save us thinking too much. Look at Fig. 15.21. You will see that at A we have three things all at right angles to each other. They are

(i) the lines of force from the north to the south pole of the magnet;
(ii) the conventional current direction in the wire;

(iii) the direction in which the wire was made to move.

No matter which way we arrange our apparatus, the thumb and first two fingers of your left hand, held each one at right angles to the other as shown in Fig. 15.22 will help you to predict the

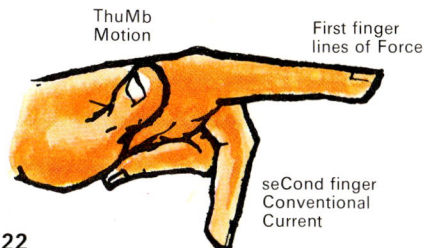
ThuMb Motion
First finger lines of Force
seCond finger Conventional Current

Fig. 15.22

direction of the force produced in this experiment. If the first finger points in the direction of the lines of force, and the second finger in the direction of the conventional current, then the thumb will be in the direction of motion of the wire (and therefore in the direction of the force produced). This can easily be remembered if you notice that the words 'First' and 'lines of Force' have f's in them; the word 'seCond' and 'Current' have c's in them; and the words 'thuMb' and 'Motion' both have m's.

15.16 SIMPLE ELECTRIC MOTORS

If a wire carrying a current moves when it is placed in a magnetic field it ought to be possible to use this fact to make a motor. As was mentioned in section 15.14 this is, in fact, the basis of all electric motors, and we are now going to make a model of one.

Experiment 15.19
Making an electric motor

Look at Fig. 15.23. The wooden block with the groove on two sides is going to be the part of the motor that spins round. This is generally called the **armature**. There are short lengths of aluminium tubing at each end of the block, and these are going to form the axle on which it rotates. One of these pieces of tubing should have a piece of Sellotape (which is an insulating material) wound round it. Cut two thin slices of valve tubing rubber to make two small rubber rings and slide these on to the tube.
Bare about 2 cm of insulation from the end of a coil of PVC-covered copper wire, loop the

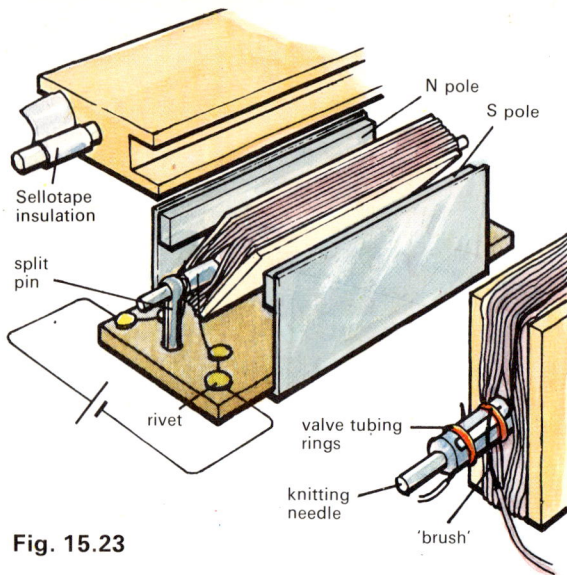

Fig. 15.23

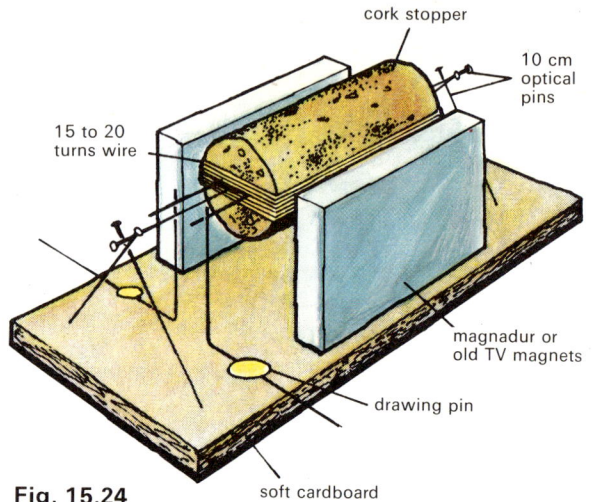

Fig. 15.24

bared end as shown, and fix it under the rubber rings. Now wind between 10 and 16 turns of wire in the grooves round the block, and cut off the wire leaving enough for about 2 cm to be bared and fixed under the rubber rings on the opposite side of the insulated aluminium tube. A knitting needle is threaded through the aluminium tube and through two split pins which form bearings so that the coil on the block can rotate. The bared copper loops of wire are going to form part of a rotating switch which in the motor is called the **commutator**.

Two wires, labelled 'brush' in the diagram, just press against the bared loops of wire on the commutator, and will be connected to a low voltage supply of current. Put the board with its assembled parts between two magnets, as shown. When the current is switched on the coil will continue to spin once you have set it in motion. If it does not work when you spin the coil one way, try setting it off in the opposite direction.

The direction of spin will depend on the direction of the current flow in the armature. Why?

Another way of making a model electric motor using a cork and pins is shown in Fig. 15.24, and you should be able to make it up from the diagram.

Make a list of all the appliances you can think of which use electric motors. Some of you will probably be able to bring old toy electric motors to school. See if you can recognize the armature (the coil), the brushes (often copper or carbon blocks), and the magnet. The faces of the magnets are often called 'pole pieces'. If your school has an old car starter motor, have a look at it. You will

find that its magnet can be made very strong. How is this done?

15.17 A METER FOR MEASURING CURRENT

It is easy to show that the bigger the current passing through a wire, the larger will be the force acting on it when it is in a magnetic field. In the electric motor we are transforming electrical energy into movement energy, and if we want to get more energy out of the motor we shall have to put more electrical energy in, that is, we shall have to increase the current flowing. We cannot do this just by increasing the voltage of the supply we put on to the motor because the increased current flowing through the wire on the armature would cause it to heat up. What, then, would you have to do to make a more powerful motor?

Since, as we have said, the force acting on a current-carrying wire in a magnetic field depends on the strength of the current, we should be able to use this fact to measure the strength of the current, and this is one way of making a current meter.

Experiment 15.20
Making a model current meter

This is very similar to the electric motor you made in Experiment 15.19 except that the armature is prevented from spinning right round by a spring. Fig. 15.25 shows you what you have to do. The drinking straw acts as a pointer and tells you how far the coil turns round.

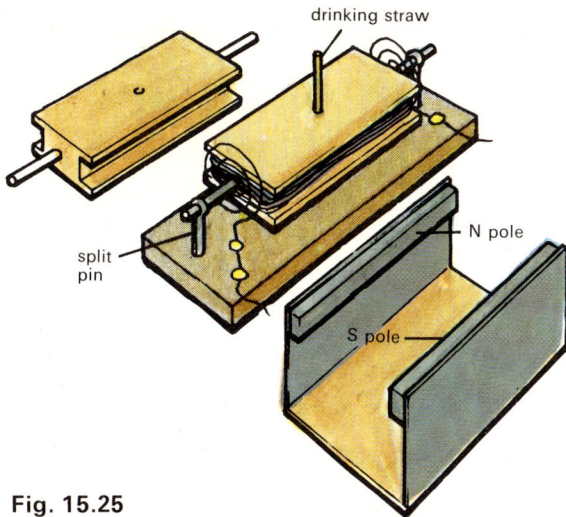

drinking straw

N pole

S pole

split pin

Fig. 15.25

Meters working on this principle are called moving-coil meters. You will have some in your laboratory. Some of the electrical meters on the dashboard of a car are of this type. Can a moving-coil meter be used to measure alternating current?

15.18 THE ELECTRICITY SUPPLY

In our electrical experiments we have used batteries, or power packs which have been connected to the mains. Where does the electrical energy in the mains supply come from? In the experiments

Fig. 15.26 A group of eminent nineteenth century scientists. *From left to right*: Faraday, Huxley, Wheatstone, Sir David Brewster, and Professor Tyndall. What was each of them famous for?

with electric motors we were using electrical energy to make mechanical or movement energy. It ought to be possible to reverse this process and make electrical energy from mechanical energy. This was first done by Michael Faraday in the early part of last century. Try to find out about this very famous scientist from a book in your library. In the next few experiments you will be repeating the kind of work he did.

Experiment 15.21
Electrical energy from magnetism

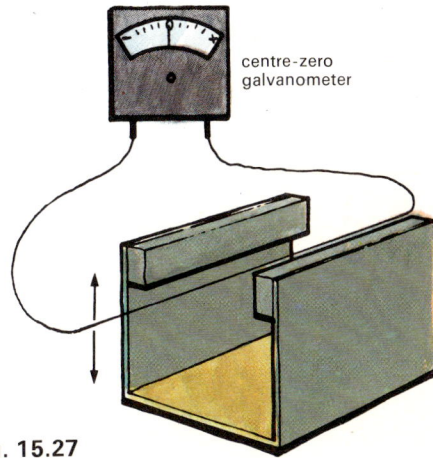

centre-zero galvanometer

Fig. 15.27

Set up the apparatus shown in Fig. 15.27. Move the wire between the poles of the permanent magnets as shown. What effect does this have on the meter? Now try moving a coil of several turns of wire over the magnet (Fig. 15.28). Is the meter reading the same as before?

Fig. 15.28

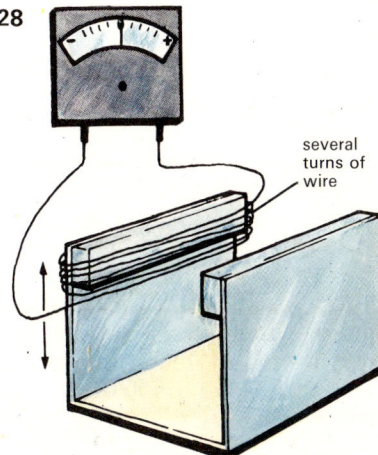

several turns of wire

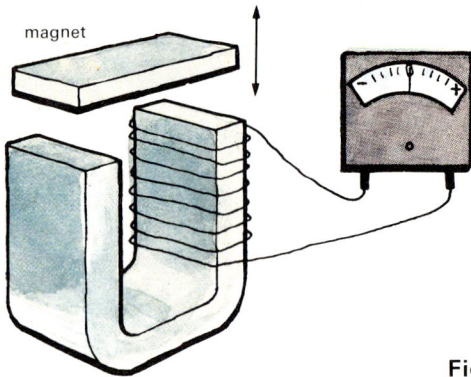

Fig. 15.29

spring

bar
magnet

Fig. 15.31

Set up a C-core with a coil of wire over one leg of it connected to a galvanometer (Fig. 15.29). Place a magnet across the ends of the core, and then remove it again. Watch the galvanometer as you do this. In which of these experiments did you make the biggest current?

If you wanted to make electricity this way it would not be very convenient for you to keep on pushing a magnet on and off a coil for a long time. You would, of course, be using up your energy, moving your hand backwards and forwards. This is energy of movement, or kinetic energy, and you are converting it into electrical energy. You have successfully performed an experiment which it took Faraday a long time to do because he did not have such a good way of measuring electric currents as you have today. He used a current detector like the one shown in Fig. 15.30. You can easily make one of these, and

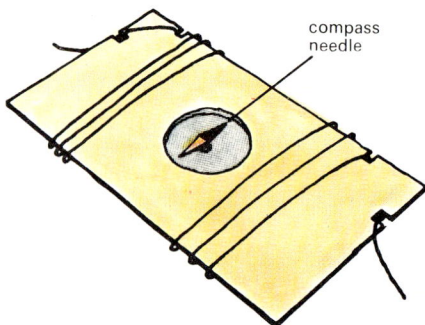

compass
needle

Fig. 15.30

can see whether it will detect the currents you make when you do Experiment 15.21. Of course, we might be able to fix up some way of moving a magnet in and out of a coil by machine instead of doing it ourselves. Here is one way of doing it.

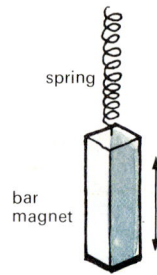

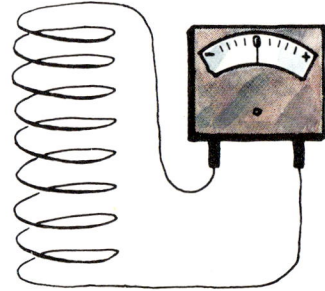

Experiment 15.22

Here we use a spring to make the magnet move in and out of the coil. Set up the apparatus shown in Fig. 15.31. What do you notice about the movement of the needle in the galvanometer as the magnet enters and withdraws from the coil?

Because the needle kicks from side to side we can say that the current in the coil is flowing first in one direction and then in the reverse. This kind of current is called alternating current. We have already referred to this on page 107. Our household electrical supply is alternating current, changing direction 100 times a second.

In all the experiments in which you have made electrical energy from a magnet and some wire you have had to move the one relative to the other. Try Experiment 15.22 again but move the coil up and down at the same rate as the magnet moves, with the magnet inside the coil all the time. Is any current generated now?

To generate a current, then, the invisible lines of force of the magnet have to be 'cut' by the turns of wire on the coil, and the current will only flow when there is movement so that the lines are cut. The earth behaves as a big magnet with lines of force running in a north–south direction. If you cut these lines of force with a wire you can generate a current. Spin a big skipping rope made of 'flex' in the playground (Fig. 15.32). In which direction will you have to spin it to cut as many lines of force as possible? The 'flex' should be

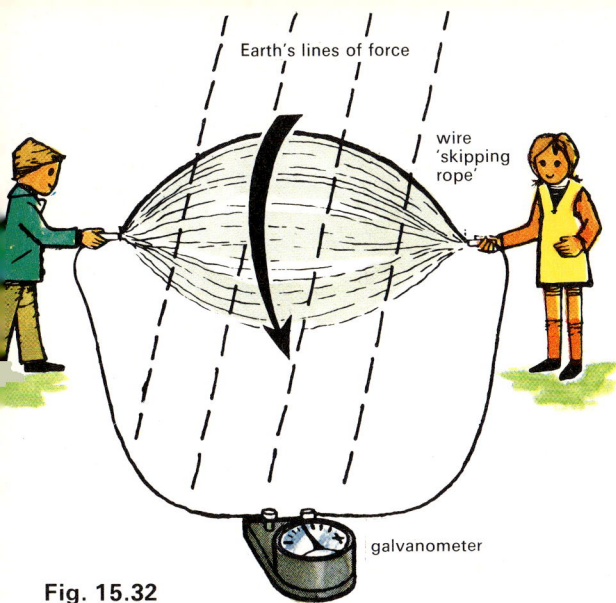

Earth's lines of force

wire 'skipping rope'

galvanometer

Fig. 15.32

connected to a galvanometer. See whether any current flows.

When we make a current in a coil, or a wire, in this way by cutting lines of force, we say that a current has been 'induced' in the wire. The 'flex' skipping rope is really the idea behind the dynamo, or the generator. All that happens in a dynamo is that a coil of wire is spun round in a magnetic field provided either by permanent magnets or by electromagnets. You will see that this is exactly the same arrangement as you had in the electric motor. In the latter you put in electrical energy and got out mechanical energy; in the dynamo you put in mechanical energy and get out electrical energy.

Experiment 15.23
The dynamo

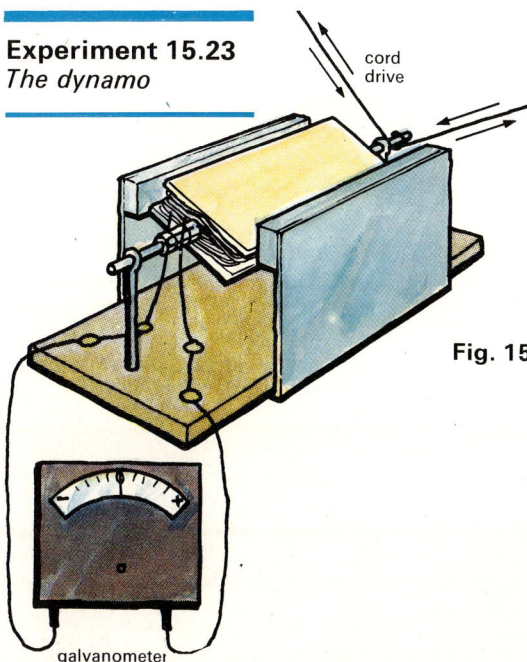

cord drive

Fig. 15.33

galvanometer

Take one of the electric motors you made in Experiment 15.19. Instead of connecting a source of electricity across it, connect it to a galvanometer. Turn the motor by winding a cord round the axle (Fig. 15.33). Is any current generated?

Experiment 15.24
The cycle dynamo

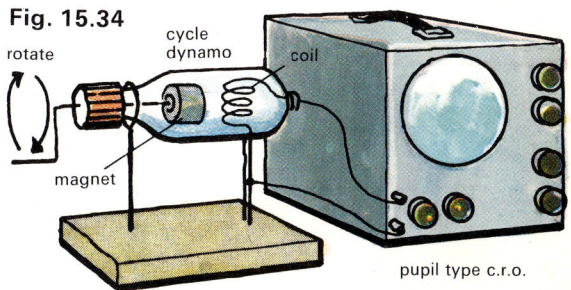

Fig. 15.34

rotate

cycle dynamo

coil

magnet

pupil type c.r.o.

Connect a cycle dynamo to a galvanometer and turn the handle (Fig. 15.34). It is interesting, too, to connect the dynamo to a cathode ray oscilloscope. If the time base is switched off, what happens to the spot on the screen when you turn the dynamo? What does this mean?

Now switch on the time base. What kind of pattern is produced as you drive the dynamo? What does it tell you about the current generated by the dynamo?

The large dynamos in our power stations are simply much bigger versions of these cycle dynamos, but they use electromagnets instead of permanent magnets. Why? They are driven in various ways. In Scotland there are several hydro-electric power stations, where the energy necessary to drive generators is obtained from water falling from a high to a low level through turbines. In other places, generators are turned by steam turbines in which the steam is raised by burning coal or oil.

You will remember that in Unit 3 (Book 1) you studied some of the ways in which energy could be converted from one form to another, and you set up several models showing how electrical energy could be made from mechanical energy. It is worth while setting up these models again. Now that you know more about it you will be able to appreciate what goes on in these energy conversions rather better.

Something for you to find out
Many diesel-electric engines are used on our railways. How do they work?

WHAT YOU HAVE LEARNT IN THIS UNIT

1. The domestic electricity supply is at a much higher voltage than that of an ordinary battery. It is therefore capable of giving you a severe 'shock' and you must always take great care not to touch any conductor connected to the mains. All such conductors are protected by a layer of insulating material through which the current cannot pass. This may be plastic or rubber. Of course, this insulation has to be removed where wires are connected inside plugs, lampholders, etc. and you must be careful never to touch these bare ends if they are connected to the mains supply.

2. After entering the house, the mains supply branches off into a number of parallel circuits. This allows each appliance to be switched on and off independently of each other, and to operate at the same voltage.

3. The power of an electrical appliance is the rate at which it uses electrical energy and is measured in joules per second, or watts. To find the power in watts multiply voltage (volts) by the current taken (amperes).

4. The electrical energy that an appliance uses is measured in joules or in kilowatt hours.

$$1 \text{ kilowatt hour} = 3\ 600\ 000 \text{ joules}$$
$$(\text{or } 3.6 \times 10^6 \text{ joules})$$

To find the number of kilowatt hours (or units) an appliance uses, multiply the power in kilowatts by the time in hours for which it is used.

5. It is important in choosing fuse wire to select the wire suitable for a current *just* above the current which the appliance normally uses.

6. The thermionic valve allows electrons to flow in one direction only – from the filament (or cathode) to the positively charged anode. These valves often need dangerously high voltages on the anode to pull the electrons away from the filament, and they produce heat. Transistors can replace thermionic valves. They need only low voltages, and in their case heat can do damage.

7. In the cathode ray oscilloscope a beam of electrons causes the screen to glow. The time base control causes this spot of light to fly backwards and forwards across the screen. The same principle is used in a television set.

8. Fluorescent lamps, used in strip lighting, are about $2\frac{1}{2}$ times as efficient as incandescent filament lamps.

9. An electric current flowing through a wire produces a magnetic effect. This effect is stronger when (i) the current is stronger; (ii) the wire is coiled; (iii) more turns are used on the coil; (iv) an iron core is put in the coil.

10. The iron core of an electromagnet is magnetized only while current flows in the coil.

11. The region round a magnet where it has an effect is called the magnetic field.

12. When an electric current flows in a wire which is in a magnetic field, the wire receives a force. Its direction is given by the left-hand rule.

13. This force on a conductor is used in electric motors, and in moving-coil meters.

14. When a wire cuts the lines of force of a magnet an electric current is induced in the coil. This effect is used in the dynamo, which makes it possible to generate electricity by using mechanical energy.

Fig. 15.35 The interior of a modern power station showing the generators

Fig. 15.36 A diesel-electric locomotive used on the London to Scotland run

Index